Ergonomic checkpoints

Ergonomic checkpoints

Practical and easy-to-implement solutions for improving safety, health and working conditions

Prepared by the International Labour Office
in collaboration with the International Ergonomics Association

International Labour Office Geneva

ILO
Ergonomic checkpoints: Practical and easy-to-implement solutions for improving safety, health and working conditions
Geneva, International Labour Office, 1996

/Guide/, /Ergonomics/, /Occupational safety/, /Occupational health/, /Work environment/.
13.04.1
ISBN 92-2-109442-1

ILO Cataloguing in Publication Data

Printed in Switzerland PUB

Foreword

Consistently high and increasing numbers of occupational accidents and diseases continue to stem from a lack of attention to ergonomics in the workplace. Much more focus has been placed on research and high technology than on practical action in the places where most people work. To date, the application of ergonomic principles has still reached only a limited number of workplaces despite its very great potential for improving working conditions and productivity. As a result, the gaps remain large in applying ergonomics to workplaces in different sectors and countries, as shown in many reports on occupational accidents, work-related diseases, major industrial accidents and unsatisfactory working conditions. *Ergonomic checkpoints* has been developed with the objective of bridging some of those gaps, particularly for small and medium-sized enterprises, by offering practical, low-cost solutions to ergonomic problems.

Many examples exist worldwide of practical, locally based ergonomic improvements including ergonomically designed hand tools, carts, materials-handling techniques, workstation arrangements, worksite welfare facilities and group work methods, in addition to the ergonomics applications developed by qualified specialists or well-trained practitioners. It is increasingly apparent that local improvements achieved at the grass-roots level should be made known to other workplaces where similar improvements are possible. We hope that using *Ergonomic checkpoints* will stimulate this valuable sharing of experiences and help to promote a more systematic application of ergonomic principles.

This manual presents 128 ergonomic interventions aimed at creating positive effects without relying on costly and highly sophisticated solutions, emphasizes realistic solutions which can be applied in a flexible manner and contributes to improved working conditions and higher productivity. It is our hope that *Ergonomic checkpoints* will inspire managers, supervisors, workers, trainers and educators, as well as ergonomics specialists, to help in the sharing of practical information and experience by filtering and disseminating ergonomically sound workplace improvements.

Dr. Chandra Pinnagoda,
Chief,
Occupational Safety and Health Branch

Preface

This book is a compilation of "ergonomic checkpoints" that can be used to find practical solutions for improving working conditions from an ergonomic point of view. Its aim is to provide a useful tool for all those who intend to improve their working conditions for better safety, health and efficiency.

Ergonomic checkpoints is intended for those wishing to apply practical improvements to existing working conditions. The book covers all the main ergonomic issues at the workplace: materials storage and handling; hand tools; productive machine safety; improving workstation design; lighting; premises; control of hazardous substances and agents; welfare facilities; and work organization. It is thus suited to identifying practical solutions to ergonomic problems in each local situation. The manual can help users to look for such solutions, as each checkpoint indicates an action. Available options for that action, as well as some additional hints, are given. Users can select those checkpoints which are applicable to their own workplace and use the "action" sentences immediately as a checklist adapted to the particular workplace. The users of this book may thus use all or part of the checkpoints, as determined by the local situation (see "Suggestions for using the manual", pp. xi-xii, for further details).

Ergonomic checkpoints is presented as the next step to the ILO publication *Higher productivity and a better place to work: Action manual*, published in 1988 as a guide to improving workplaces in small and medium-sized enterprises. This earlier publication has been used widely in training activities in different regions of the world. The manual is an effective tool in the ILO's training approach, "Work Improvement in Small Enterprises" (WISE). The approach is being applied to a variety of activities in many developing countries within the framework of the ILO's International Programme for the Improvement of Working Conditions and Environment (known as PIACT after its French name). As a follow-up to these activities, *Ergonomic checkpoints* forms an integral part of the ILO programme.

Ergonomic checkpoints is the result of collaboration between the International Labour Office and the International Ergonomics Association (IEA). In 1991, the IEA Technology Transfer Committee established a group of experts, chaired by Najmedin Meshkati, to create an outline of the document and to produce a major part of the material. The group was piloted mainly by Kazutaka Kogi from the ILO and Ilkka Kuorinka from the IEA. Tuulikki Kuorinka assembled the different manuscripts and Kazutaka Kogi further edited the checkpoints.

The core group for compiling *Ergonomic checkpoints* consisted of the following people:

- Martin Helander, State University of New York, Buffalo, United States;
- Andrew Imada, University of Southern California, Los Angeles, United States;
- Kazutaka Kogi,* International Labour Office, Geneva, Switzerland;
- Stephen Konz, Kansas State University, Manhattan, United States;
- Ilkka Kuorinka, Institut de Recherche en Santé et en Sécurité de Travail de Québec (IRSST), Montreal, Canada;
- Tuulikki Kuorinka, IRSST, Montreal, Canada;
- Wolfgang Laurig, Institut für Arbeitsphysiologie, Dortmund, Germany;
- Najmedin Meshkati, University of Southern California, Los Angeles, United States;
- Houshang Shahnavaz, Luleå University of Technology, Luleå, Sweden.

The group of experts identified several main areas where the contribution of ergonomics to working conditions was assessed to be most important for small enterprises. For each area, 10 to 20 checkpoint items were developed. As a result, 128 checkpoints have been put together. In developing the checkpoints, the emphasis has been on a concrete and visual presentation, with the goals of problem solving and solution finding. The analytical part, therefore, is minimized in favour of practical solutions. This is concordant with the approach used

* Presently Director of the Institute for Science of Labour, Kawasaki, Japan.

in the ILO programme that has proved to be effective in both industrially developed and developing countries.

In 1993, the checkpoints were tested in Indonesia and Thailand in two "roving seminars" organized jointly by the ILO and the IEA, in collaboration with the South-East Asian Ergonomics Society. These seminars demonstrated that the checkpoints can be used effectively in training local people to find realistic solutions to ergonomic problems at workplaces in a developing situation.

The manual is a product of the joint efforts of many people who worked together with the group of experts. Hamid Kavianian, California State University, Long Beach, and Karl Kroemer, Virginia Polytechnic Institute and State University, Radford, United States, took part in the drafting of some of the checkpoints. Jürgen Serbitzer and Valentina Forastieri of the ILO conducted initial research.

Ellen Rosskam Krasnosselski and Juan-Carlos Hiba of the ILO compiled the illustrations together with the expert group. They include graphics drawn by Vlad Ganea and Igor Lossavio. Other drawings are taken from *Safety, health and working conditions*, a training manual published by the Joint Industrial Safety Council of Sweden (Stockholm, 1987), and from previous ILO publications.

At the request of the IEA Education and Training Committee (Chair, Margaret Bullock), the following people worked together in applying the checkpoints in the "roving seminars": Kamiel Vanwonterghem, KV-Ergonomics, Hasselt, Belgium; Kitti Intaranont, Chulalongkorn University, Bangkok, Thailand; Chaiyuth Chavalitnitikul, Ministry of Labour and Social Welfare, National Institute for the Improvement of Working Conditions and Environment, Bangkok, Thailand; and Adnyana Manuaba, University of Udayana, Denpasar, Indonesia. The collaboration of the IEA led by Hal W. Hendrick, former President, is highly appreciated. Thanks are due also to the IEA Council for both material and intellectual support. The initiative of the ILO in realizing this project as part of its regular budget and programme activities is also greatly appreciated. Thanks are particularly due to Claude Dumont, Director of the Working Conditions and Environment Department, and Chandra Pinnagoda, Chief of the Occupational Safety and Health Branch, for their support.

The authors hope that this new publication will serve as an impetus for practical improvements at many places of work in different parts of the world.

Contents

Suggestions for using the manual

The suggestions given here for using *Ergonomic checkpoints* are based on the way in which the manual was used in the pilot test "roving seminars" organized by the ILO. We believe that there are a wide variety of ways in which this book can be used and that the creativity of the instructor will determine the flexibility of the manual.

In implementing workplace improvements, the checkpoints provide guidance based on a number of underlying principles, which include the following:

- immediate solutions need to be developed with the active involvement of employers, and the support of workers must be mobilized;
- group work is advantageous for planning and implementing practical improvements;
- the use of available local material and expertise has many benefits;
- multifaceted action should ensure that improvements are sustained over time; and
- continuing action programmes are needed to create improvements.

A variety of people can use this book when checking existing workplace conditions or when examining workplace plans at the design stage. At the same time, users can learn various kinds of solutions that are simple, inexpensive and therefore applicable even in small workplaces. The checkpoints have been developed for use by employers, supervisors, workers, engineers, safety and health personnel, trainers and educators, inspectors, extension workers, ergonomists, workplace designers and other people who are interested in improving workplaces, work equipment and working conditions.

Ergonomic checkpoints is intended for those who wish to improve working conditions by means of an organized analysis and who look for practical solutions to their own local problems. For these purposes the manual covers all the main ergonomic issues at the workplace. Covering the principal issues will help users to check their workplaces in an organized manner. Users who would like to know about available solutions to their particular problem may go straight to the specific checkpoint which addresses it, because the checkpoints are particularly suited to looking at various workplaces and identifying practical solutions applicable in each local situation.

Another unique feature of the ergonomic checkpoints is the way in which they are presented. Each checkpoint indicates an action. Options for that action are then described. When the titles of all or part of the checkpoints are put together in the form of a list, this can be used immediately as a checklist. A sample checklist comprising all 128 checkpoints is included in this manual. Users may use all or part of the checklist, as determined by the local situation.

The following are the best ways to use *Ergonomic checkpoints* for the purpose of improving a workplace:

1. Use the checklist on pages xiii-xxii of the manual to select and apply those checkpoints which are relevant to the user's own workplace. The selected checkpoints then become the user's own checklist.
2. Organize a group discussion using the user's own checklist as guidance material.
3. A group of people can check the workplace by conducting a walk-through survey applying their own checklist.
4. A group discussion should follow the walk-through survey combined with the application of the user's own checklist. This should concentrate on determining the priority improvements to be proposed for immediate action.
5. During the group discussion, the "how" information and "some more hints" in this manual may be useful as additional information for the selected checkpoints.
6. Good working conditions and good work practices should be mentioned where they were observed.

The use of such a locally adapted checklist made up from selected checkpoints can be a powerful tool for a training course on ergonomic workplace improvements. For example, in the "roving seminars" organized in Thailand and Indonesia as pilot courses for using *Ergonomic checkpoints*, a 44-item checklist was used. The checklist comprised items selected from sections on materials handling, workstation design, machine safety, premises, welfare facilities and work organization.

The participants of the seminar first took part in a checklist exercise in which they visited a factory and used the 44-item checklist to identify improvements. Using the parts of the manual corresponding to the 44 items, they then discussed in small groups what priority actions and practical improvements might be proposed jointly to the factory management. At the end of each session, the results of group discussions were presented to all the participants together. Before holding a final session, the participants discussed, again in small groups, the final priority suggestions for improving the workstations in the factory. These proposals were presented to the factory managers who joined in the final session. Similar training courses using an ergonomics checklist as a tool for group discussion have proved very effective.

Users of the manual are likely, through group discussion, to find that there is wide room for improvement, even in "good workplaces". In our "roving seminars", managers of the workplaces where the walk-through surveys were conducted were invited to attend discussions where the priority improvements recommended by the users were presented. In our experience, the practical, inexpensive and locally applicable recommendations for making workplace improvements were appreciated by management, particularly in small and medium-sized enterprises.

Instructors using the manual for training purposes may wish to make transparencies from the illustrations, for use with an overhead projector, where this is available.

In "train-the-trainer courses", it is both useful and interesting to encourage several trainees to introduce and use the checkpoints as a way to familiarize the other participants with the concept of ergonomics.

Your feedback is important to us and we would like to hear from you. Please send your comments on the manual and on the ways in which you have used it to Dr. C. Pinnagoda, Chief, Occupational Safety and Health Branch, International Labour Office (ILO), 4 route des Morillons, CH-1211 Geneva 22, Switzerland, or send us a fax at + 41 022 7996878.

ERGONOMICS CHECKLIST

How to use the checklist

1. Ask the manager any questions you have. You should learn about the main products and production methods, the number of workers (male and female), the hours of work (including breaks and overtime) and any important labour problems.
2. Define the work area to be checked. In the case of a small enterprise the whole production area can be checked. In the case of a larger enterprise, particular work areas can be defined for separate checking.
3. Read through the checklist and spend a few minutes walking around the work area before starting to check.
4. Read each item carefully. Look for a way to apply the measure. If necessary, ask the manager or workers questions. If the measure has already been applied or it is not needed, mark NO under "Do you propose action?" If you think the measure would be worth while, mark YES. Use the space under REMARKS to put a description of your suggestion or its location.
5. After you have finished, look again at the items you have marked YES. Choose a few where the benefits seem likely to be the most important. Mark PRIORITY for these items.
6. Before finishing, make sure that for each item you have marked either NO or YES, and that for some items marked YES you have marked PRIORITY.

Materials storage and handling

1. Clear and mark transport routes.

   Do you propose action?

   ☐ No ☐ Yes ☐ Priority

   Remarks ______________________________

   ____

2. Keep aisles and corridors wide enough to allow two-way transport.

   Do you propose action?

   ☐ No ☐ Yes ☐ Priority

   Remarks ______________________________

   ____

3. Make the surface of transport routes even, not slippery and without obstacles.

   Do you propose action?

   ☐ No ☐ Yes ☐ Priority

   Remarks ______________________________

   ____

4. Provide ramps with a small inclination of up to 5-8 per cent instead of small stairways or sudden height differences within the workplace.

   Do you propose action?

   ☐ No ☐ Yes ☐ Priority

   Remarks ______________________________

   ____

5. Improve the layout of the work area so that the need to move materials is minimized.

   Do you propose action?

   ☐ No ☐ Yes ☐ Priority

   Remarks ______________________________

   ____

6. Use carts, hand-trucks and other wheeled devices or rollers, when moving materials.

   Do you propose action?

   ☐ No ☐ Yes ☐ Priority

   Remarks ______________________________

   ____

7. Use mobile storage racks to avoid unnecessary loading and unloading.

   Do you propose action?

   ☐ No ☐ Yes ☐ Priority

   Remarks ______________________________

   ____

8. Use multi-level shelves or racks near the work area in order to minimize manual transport of materials.

   Do you propose action?

   ☐ No ☐ Yes ☐ Priority

   Remarks ______________________________

   ____

9. Use mechanical devices for lifting, lowering and moving heavy materials.

   Do you propose action?

   ☐ No ☐ Yes ☐ Priority

   Remarks ______________________

   ____

10. Reduce manual handling of materials by using conveyers, hoists and other mechanical means of transport.

   Do you propose action?

   ☐ No ☐ Yes ☐ Priority

   Remarks ______________________

   ____

11. Instead of carrying heavy weights, divide them into smaller lightweight packages, containers or trays.

   Do you propose action?

   ☐ No ☐ Yes ☐ Priority

   Remarks ______________________

   ____

12. Provide handholds, grips or good holding points for all packages and containers.

   Do you propose action?

   ☐ No ☐ Yes ☐ Priority

   Remarks ______________________

   ____

13. Eliminate or minimize height differences when materials are moved manually.

   Do you propose action?

   ☐ No ☐ Yes ☐ Priority

   Remarks ______________________

   ____

14. Feed and remove heavy materials horizontally by pushing and pulling them instead of raising and lowering them.

   Do you propose action?

   ☐ No ☐ Yes ☐ Priority

   Remarks ______________________

   ____

15. Eliminate tasks that require bending or twisting while handling materials.

   Do you propose action?

   ☐ No ☐ Yes ☐ Priority

   Remarks ______________________

   ____

16. Keep objects close to the body when carrying.

   Do you propose action?

   ☐ No ☐ Yes ☐ Priority

   Remarks ______________________

   ____

17. Raise and lower materials slowly in front of the body without twisting or deep bending.

   Do you propose action?

   ☐ No ☐ Yes ☐ Priority

   Remarks ______________________

   ____

18. When carrying a load for more than a short distance, spread the load evenly across the shoulders to provide balance and reduce effort.

   Do you propose action?

   ☐ No ☐ Yes ☐ Priority

   Remarks ______________________

   ____

19. Combine heavy lifting with physically lighter tasks to avoid injury and fatigue and to increase efficiency.

   Do you propose action?

   ☐ No ☐ Yes ☐ Priority

   Remarks ______________________

   ____

20. Provide conveniently placed waste containers.

   Do you propose action?

   ☐ No ☐ Yes ☐ Priority

   Remarks ______________________

   ____

21. Mark escape routes and keep them clear of obstacles.

   Do you propose action?

   ☐ No ☐ Yes ☐ Priority

   Remarks ______________________

   ____

Hand tools

22. Use special-purpose tools for repeated tasks.

    Do you propose action?

    ☐ No ☐ Yes ☐ Priority

    Remarks ______________________

23. Provide safe power tools and make sure that safety guards are used.

    Do you propose action?

    ☐ No ☐ Yes ☐ Priority

    Remarks ______________________

24. Use hanging tools for operations repeated in the same place.

    Do you propose action?

    ☐ No ☐ Yes ☐ Priority

    Remarks ______________________

25. Use vices and clamps to hold materials or work items.

    Do you propose action?

    ☐ No ☐ Yes ☐ Priority

    Remarks ______________________

26. Provide hand support when using precision tools.

    Do you propose action?

    ☐ No ☐ Yes ☐ Priority

    Remarks ______________________

27. Minimize the weight of tools (except for striking tools).

    Do you propose action?

    ☐ No ☐ Yes ☐ Priority

    Remarks ______________________

28. Choose tools that can be operated with minimum force.

    Do you propose action?

    ☐ No ☐ Yes ☐ Priority

    Remarks ______________________

29. For hand tools, provide the tool with a grip of the proper thickness, length and shape for easy handling.

    Do you propose action?

    ☐ No ☐ Yes ☐ Priority

    Remarks ______________________

30. Provide hand tools with grips that have adequate friction or with guards or stoppers to avoid slips and pinches.

    Do you propose action?

    ☐ No ☐ Yes ☐ Priority

    Remarks ______________________

31. Provide tools with proper insulation to avoid burns and electric shocks.

    Do you propose action?

    ☐ No ☐ Yes ☐ Priority

    Remarks ______________________

32. Minimize vibration and noise of hand tools.

    Do you propose action?

    ☐ No ☐ Yes ☐ Priority

    Remarks ______________________

33. Provide a "home" for each tool.

    Do you propose action?

    ☐ No ☐ Yes ☐ Priority

    Remarks ______________________

34. Inspect and maintain hand tools regularly.

    Do you propose action?

    ☐ No ☐ Yes ☐ Priority

    Remarks ______________________

35. Train workers before allowing them to use power tools.

    Do you propose action?

    ☐ No ☐ Yes ☐ Priority

    Remarks ______________________

36. Provide for enough space and stable footing for power tool operation.

    Do you propose action?

    ☐ No ☐ Yes ☐ Priority

    Remarks ____________________

Productive machine safety

37. Protect controls to prevent accidental activation.

    Do you propose action?

    ☐ No ☐ Yes ☐ Priority

    Remarks

38. Make emergency controls clearly visible and easily accessible from the natural position of the operator.

    Do you propose action?

    ☐ No ☐ Yes ☐ Priority

    Remarks

39. Make different controls easy to distinguish from each other.

    Do you propose action?

    ☐ No ☐ Yes ☐ Priority

    Remarks

40. Make sure that the worker can see and reach all controls comfortably.

    Do you propose action?

    ☐ No ☐ Yes ☐ Priority

    Remarks

41. Locate controls in sequence of operation.

    Do you propose action?

    ☐ No ☐ Yes ☐ Priority

    Remarks

42. Use natural expectations for control movements.

    Do you propose action?

    ☐ No ☐ Yes ☐ Priority

    Remarks

43. Limit the number of foot pedals and, if used, make them easy to operate.

    Do you propose action?

    ☐ No ☐ Yes ☐ Priority

    Remarks

44. Make displays and signals easy to distinguish from each other and easy to read.

    Do you propose action?

    ☐ No ☐ Yes ☐ Priority

    Remarks

45. Use markings or colours on displays to help workers understand what to do.

    Do you propose action?

    ☐ No ☐ Yes ☐ Priority

    Remarks

46. Remove or cover all unused displays.

    Do you propose action?

    ☐ No ☐ Yes ☐ Priority

    Remarks

47. Use symbols only if they are easily understood by local people.

    Do you propose action?

    ☐ No ☐ Yes ☐ Priority

    Remarks

48. Make labels and signs easy to see, easy to read and easy to understand.

    Do you propose action?

    ☐ No ☐ Yes ☐ Priority

    Remarks

49. Use warning signs that workers understand easily and correctly.

    Do you propose action?

    ☐ No ☐ Yes ☐ Priority

    Remarks ______________________

50. Use jigs and fixtures to make machine operation stable, safe and efficient.

    Do you propose action?

    ☐ No ☐ Yes ☐ Priority

    Remarks ______________________

51. Purchase safe machines.

    Do you propose action?

    ☐ No ☐ Yes ☐ Priority

    Remarks ______________________

52. Use feeding and ejection devices to keep the hands away from dangerous parts of machinery.

    Do you propose action?

    ☐ No ☐ Yes ☐ Priority

    Remarks ______________________

53. Use properly fixed guards or barriers to prevent contact with moving parts of machines.

    Do you propose action?

    ☐ No ☐ Yes ☐ Priority

    Remarks ______________________

54. Use interlock barriers to make it impossible for workers to reach dangerous points when the machine is in operation.

    Do you propose action?

    ☐ No ☐ Yes ☐ Priority

    Remarks ______________________

55. Inspect, clean and maintain machines regularly, including electric wiring.

    Do you propose action?

    ☐ No ☐ Yes ☐ Priority

    Remarks ______________________

56. Train workers for safe and efficient operation.

    Do you propose action?

    ☐ No ☐ Yes ☐ Priority

    Remarks ______________________

Improving workstation design

57. Adjust the working height for each worker at elbow level or slightly below it.

    Do you propose action?

    ☐ No ☐ Yes ☐ Priority

    Remarks ______________________

58. Make sure that smaller workers can reach controls and materials in a natural posture.

    Do you propose action?

    ☐ No ☐ Yes ☐ Priority

    Remarks ______________________

59. Make sure that the largest worker has enough space for moving the legs and body easily.

    Do you propose action?

    ☐ No ☐ Yes ☐ Priority

    Remarks ______________________

60. Place frequently used materials, tools and controls within easy reach.

    Do you propose action?

    ☐ No ☐ Yes ☐ Priority

    Remarks ______________________

61. Provide a stable multi-purpose work surface at each workstation.

    Do you propose action?

    ☐ No ☐ Yes ☐ Priority

    Remarks ______________________

62. Provide sitting workplaces for workers performing tasks requiring precision or detailed inspection of work items, and standing workplaces for workers performing tasks requiring body movements and greater force.

    Do you propose action?

    ☐ No ☐ Yes ☐ Priority

    Remarks ____________________

63. Make sure that the workers can stand naturally, with weight on both feet, and perform work close to and in front of the body.

    Do you propose action?

    ☐ No ☐ Yes ☐ Priority

    Remarks ____________________

64. Allow workers to alternate standing and sitting at work as much as possible.

    Do you propose action?

    ☐ No ☐ Yes ☐ Priority

    Remarks ____________________

65. Provide standing workers with chairs or stools for occasional sitting.

    Do you propose action?

    ☐ No ☐ Yes ☐ Priority

    Remarks ____________________

66. Provide sitting workers with good adjustable chairs with a backrest.

    Do you propose action?

    ☐ No ☐ Yes ☐ Priority

    Remarks ____________________

67. Provide adjustable work surfaces for workers who alternate work between small and large objects.

    Do you propose action?

    ☐ No ☐ Yes ☐ Priority

    Remarks ____________________

68. Use a display-and-keyboard workstation, such as a visual display unit (VDU), that workers can adjust.

    Do you propose action?

    ☐ No ☐ Yes ☐ Priority

    Remarks ____________________

69. Provide eye examination and proper glasses for workers using a visual display regularly.

    Do you propose action?

    ☐ No ☐ Yes ☐ Priority

    Remarks ____________________

70. Provide up-to-date training for visual display unit (VDU) workers.

    Do you propose action?

    ☐ No ☐ Yes ☐ Priority

    Remarks ____________________

71. Involve workers in the improved design of their own workstation.

    Do you propose action?

    ☐ No ☐ Yes ☐ Priority

    Remarks ____________________

Lighting

72. Increase the use of daylight.

    Do you propose action?

    ☐ No ☐ Yes ☐ Priority

    Remarks ____________________

73. Use light colours for walls and ceilings when more light is needed.

    Do you propose action?

    ☐ No ☐ Yes ☐ Priority

    Remarks ____________________

74. Light up corridors, staircases, ramps and other areas where people may be.

    Do you propose action?

    ☐ No ☐ Yes ☐ Priority

    Remarks ____________________

75. Light up the work area evenly to minimize changes in brightness.

    Do you propose action?

    ☐ No ☐ Yes ☐ Priority

    Remarks ____________________

76. Provide sufficient lighting for workers so that they can work efficiently and comfortably at all times.

    Do you propose action?

    ☐ No ☐ Yes ☐ Priority

    Remarks ____________________

77. Provide local lights for precision or inspection work.

    Do you propose action?

    ☐ No ☐ Yes ☐ Priority

    Remarks ____________________

78. Relocate light sources or provide shields to eliminate direct glare.

    Do you propose action?

    ☐ No ☐ Yes ☐ Priority

    Remarks ____________________

79. Remove shiny surfaces from the worker's field of vision to eliminate indirect glare.

    Do you propose action?

    ☐ No ☐ Yes ☐ Priority

    Remarks ____________________

80. Choose an appropriate visual task background for tasks requiring close, continuous attention.

    Do you propose action?

    ☐ No ☐ Yes ☐ Priority

    Remarks ____________________

81. Clean windows and maintain light sources.

    Do you propose action?

    ☐ No ☐ Yes ☐ Priority

    Remarks ____________________

Premises

82. Protect the worker from excessive heat.

    Do you propose action?

    ☐ No ☐ Yes ☐ Priority

    Remarks ____________________

83. Protect the workplace from excessive outside heat and cold.

    Do you propose action?

    ☐ No ☐ Yes ☐ Priority

    Remarks ____________________

84. Isolate or insulate sources of heat or cold.

    Do you propose action?

    ☐ No ☐ Yes ☐ Priority

    Remarks ____________________

85. Install effective local exhaust systems which allow efficient and safe work.

    Do you propose action?

    ☐ No ☐ Yes ☐ Priority

    Remarks ____________________

86. Increase the use of natural ventilation when needed to improve the indoor climate.

    Do you propose action?

    ☐ No ☐ Yes ☐ Priority

    Remarks ____________________

87. Improve and maintain ventilation systems to ensure workplace air quality.

    Do you propose action?

    ☐ No ☐ Yes ☐ Priority

    Remarks ____________________

Control of hazardous substances and agents

88. Isolate or cover noisy machines or parts of machines.

    Do you propose action?

    ☐ No ☐ Yes ☐ Priority

    Remarks ______________________________

89. Maintain tools and machines regularly in order to reduce noise.

    Do you propose action?

    ☐ No ☐ Yes ☐ Priority

    Remarks ______________________________

90. Make sure that noise does not interfere with communication, safety and work efficiency.

    Do you propose action?

    ☐ No ☐ Yes ☐ Priority

    Remarks ______________________________

91. Reduce vibration affecting workers in order to improve safety, health and work efficiency.

    Do you propose action?

    ☐ No ☐ Yes ☐ Priority

    Remarks ______________________________

92. Choose electric hand lamps that are well insulated against electric shock and heat.

    Do you propose action?

    ☐ No ☐ Yes ☐ Priority

    Remarks ______________________________

93. Ensure safe wiring connections for equipment and lights.

    Do you propose action?

    ☐ No ☐ Yes ☐ Priority

    Remarks ______________________________

94. Protect workers from chemical risks so that they can perform their work safely and efficiently.

    Do you propose action?

    ☐ No ☐ Yes ☐ Priority

    Remarks ______________________________

Welfare facilities

95. Provide and maintain good changing, washing and sanitary facilities to ensure good hygiene and tidiness.

    Do you propose action?

    ☐ No ☐ Yes ☐ Priority

    Remarks ______________________________

96. Provide drinking facilities, eating areas and rest rooms to ensure good performance and well-being.

    Do you propose action?

    ☐ No ☐ Yes ☐ Priority

    Remarks ______________________________

97. Improve welfare facilities and services together with workers.

    Do you propose action?

    ☐ No ☐ Yes ☐ Priority

    Remarks ______________________________

98. Provide a place for workers' meetings and training.

    Do you propose action?

    ☐ No ☐ Yes ☐ Priority

    Remarks ______________________________

99. Clearly mark areas requiring the use of personal protective equipment.

    Do you propose action?

    ☐ No ☐ Yes ☐ Priority

    Remarks ______________________________

100. Provide personal protective equipment that gives adequate protection.

    Do you propose action?

    ☐ No ☐ Yes ☐ Priority

    Remarks ______________________________

101. Choose well-fitted and easy-to-maintain personal protective equipment when risks cannot be eliminated by other means.

Do you propose action?

☐ No ☐ Yes ☐ Priority

Remarks ______________________________

102. Ensure regular use of personal protective equipment by proper instructions, adaptation trials and training.

Do you propose action?

☐ No ☐ Yes ☐ Priority

Remarks ______________________________

103. Make sure that everyone uses personal protective equipment where it is needed.

Do you propose action?

☐ No ☐ Yes ☐ Priority

Remarks ______________________________

104. Make sure that personal protective equipment is acceptable to the workers.

Do you propose action?

☐ No ☐ Yes ☐ Priority

Remarks ______________________________

105. Provide support for cleaning and maintaining personal protective equipment regularly.

Do you propose action?

☐ No ☐ Yes ☐ Priority

Remarks ______________________________

106. Provide proper storage for personal protective equipment.

Do you propose action?

☐ No ☐ Yes ☐ Priority

Remarks ______________________________

107. Assign responsibility for day-to-day cleaning and housekeeping.

Do you propose action?

☐ No ☐ Yes ☐ Priority

Remarks ______________________________

Work organization

108. Involve workers in planning their day-to-day work.

Do you propose action?

☐ No ☐ Yes ☐ Priority

Remarks ______________________________

109. Consult workers on improving working-time arrangements.

Do you propose action?

☐ No ☐ Yes ☐ Priority

Remarks ______________________________

110. Solve work problems by involving workers in groups.

Do you propose action?

☐ No ☐ Yes ☐ Priority

Remarks ______________________________

111. Consult workers when there are changes in production and when improvements are needed for safer, easier and more efficient work.

Do you propose action?

☐ No ☐ Yes ☐ Priority

Remarks ______________________________

112. Reward workers for their help in improving productivity and the workplace.

Do you propose action?

☐ No ☐ Yes ☐ Priority

Remarks ______________________________

113. Inform workers frequently about the results of their work.

Do you propose action?

☐ No ☐ Yes ☐ Priority

Remarks ______________________________

114. Train workers to take responsibility and give them the means for making improvements in their jobs.

Do you propose action?

☐ No ☐ Yes ☐ Priority

Remarks ______________________________

115. Provide opportunities for easy communication and mutual support at the workplace.

Do you propose action?

☐ No ☐ Yes ☐ Priority

Remarks ______________________

116. Provide opportunities for workers to learn new skills.

Do you propose action?

☐ No ☐ Yes ☐ Priority

Remarks ______________________

117. Set up work groups, each of which collectively carries out work and is responsible for its results.

Do you propose action?

☐ No ☐ Yes ☐ Priority

Remarks ______________________

118. Improve jobs that are difficult and disliked in order to increase productivity in the long run.

Do you propose action?

☐ No ☐ Yes ☐ Priority

Remarks ______________________

119. Combine tasks to make the work more interesting and varied.

Do you propose action?

☐ No ☐ Yes ☐ Priority

Remarks ______________________

120. Set up a small stock of unfinished products (buffer stock) between different workstations.

Do you propose action?

☐ No ☐ Yes ☐ Priority

Remarks ______________________

121. Combine visual display work with other tasks to increase productivity and reduce fatigue.

Do you propose action?

☐ No ☐ Yes ☐ Priority

Remarks ______________________

122. Provide short and frequent pauses during continuous visual display work.

Do you propose action?

☐ No ☐ Yes ☐ Priority

Remarks ______________________

123. Consider workers' skills and preferences in assigning people to jobs.

Do you propose action?

☐ No ☐ Yes ☐ Priority

Remarks ______________________

124. Adapt facilities and equipment to disabled workers so that they can do their jobs safely and efficiently.

Do you propose action?

☐ No ☐ Yes ☐ Priority

Remarks ______________________

125. Give due attention to the safety and health of pregnant women.

Do you propose action?

☐ No ☐ Yes ☐ Priority

Remarks ______________________

126. Take measures so that older workers can perform work safely and efficiently.

Do you propose action?

☐ No ☐ Yes ☐ Priority

Remarks ______________________

127. Establish emergency plans to ensure correct emergency operations, easy access to facilities and rapid evacuation.

Do you propose action?

☐ No ☐ Yes ☐ Priority

Remarks ______________________

128. Learn about and share ways to improve your workplace from good examples in your own enterprise or in other enterprises.

Do you propose action?

☐ No ☐ Yes ☐ Priority

Remarks ______________________

Materials storage and handling

CHECKPOINT 1

Clear and mark transport routes.

WHY

Clear transport routes with easy access to worksites and storage areas greatly help to achieve a better work flow, as well as to ensure safe and quick transport.

If transport areas are not clearly marked, materials, work items and wastes tend to pile up on transport routes. These irregular piles not only obstruct transport and production, but also cause accidents.

Marking of transport routes is the most simple yet effective method to keep them clear.

HOW

1. Define transport routes as distinct from storage areas to worksites or between worksites. Consult workers about how to indicate necessary transport routes. Remove obstacles. Then install floor markings using paint on both edges of each transport route.

2. Where markings of transport routes are placed near moving machines or stored materials, provide fences or hand-rails to make movements of workers safe.

3. Make sure that nothing is placed or left on the defined transport routes. Cooperation of everyone in the workplace is necessary. Ensure that proper places for storage and waste disposal are present near the worksites. Insist until this practice of placing nothing on the floor is well established.

SOME MORE HINTS

— Providing storage racks, shelves or pallets should accompany the marking of transport routes. This helps to establish the practice of keeping transport routes clear of obstacles. The provision of waste receptacles is also important.

— In a workplace, there are usually central (or major) transport routes and secondary (or minor) ones. Central transport routes should be wide enough to allow busy transport work. Pay attention also to minor transport routes. Always mark all transport routes.

— Sometimes it becomes necessary to rearrange the layout of the work area, as a whole or in part, in order to have shorter and more efficient transport routes. This may require extra effort, but is worth trying.

POINTS TO REMEMBER

Marking transport routes is the starting-point for keeping them clear of obstacles. Clear transport routes ensure a good flow of materials and prevent accidents.

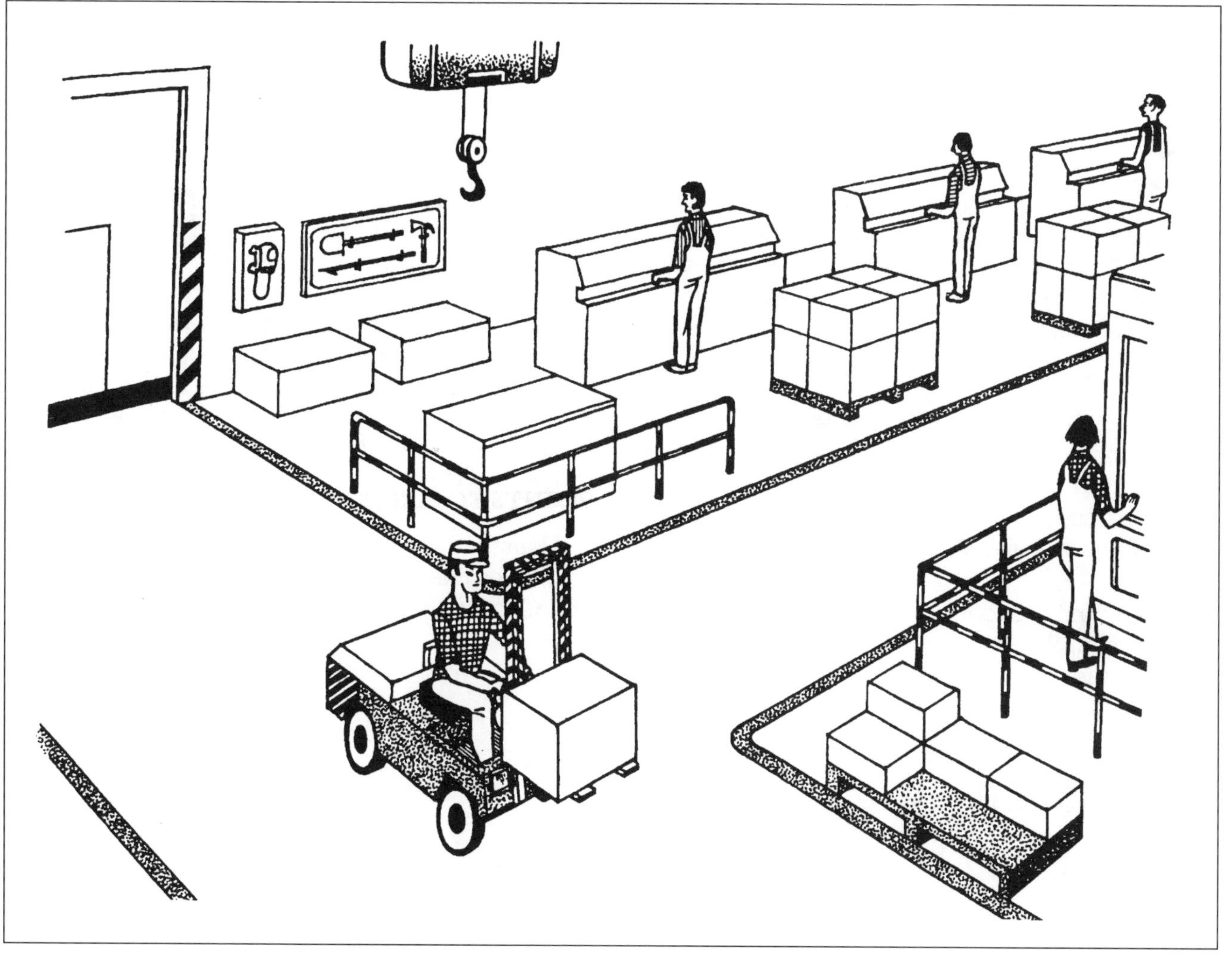

Figure 1. Draw lines to separate transport areas from working areas and keep the areas clear

CHECKPOINT 2

Keep aisles and corridors wide enough to allow two-way transport.

WHY

Aisles and corridors free from obstacles are important for the smooth movement of materials and workers. Passageways which are too narrow or which have obstacles placed in them greatly hamper the flow of work and cause considerable loss of time.

Two-way transport is a minimum requirement for any aisle or corridor. Smooth two-way transport can help good work flow and can also prevent accidents. There should be very few exceptions to this rule (e.g. dead-end corners of small storage areas which are only occasionally used).

Aisles and corridors wide enough to allow the passage of pushcarts greatly facilitate efficient production, as fewer and safer transport operations become possible.

HOW

1. Clear aisles and corridors of obstacles so that smooth passage is always possible. Mark aisles on both sides.

2. Make aisles for transport of materials wide enough (at least 125-140 cm) to allow two-way transport. Minor passageways where transport is infrequent can be at least 75 cm, but keep such exceptions to a minimum.

3. Check if mobile racks and pushcarts can easily pass through the aisles and corridors.

4. Where two-way transport is not possible (e.g. because of space constraints despite frequent transport), consider alternative easier ways to transport materials and semi-products, such as the use of easy-to-carry pallets, small trays or detachable shelves that can be placed on pushcarts after arriving at two-way aisles.

SOME MORE HINTS

- Where possible and appropriate, place fences or partitions around transport routes in order to ensure that they are always free for easy passage.

- Mobile racks or pushcarts can greatly improve the efficiency of transport. If their use is hampered by narrow aisles, do not hesitate to introduce arrangements (such as relocation of machines) to improve their smooth passage.

- Corners in passageways can create congestion. Make turning around corners smooth by allowing sufficient space.

- In order to avoid placing materials in the aisles and corridors, provide racks, side-stands and shelves so that people are more easily guided to respect the rule of not placing materials on the floor.

POINTS TO REMEMBER

Easy two-way transport through aisles and corridors saves time and energy, and helps to keep the workplace in order.

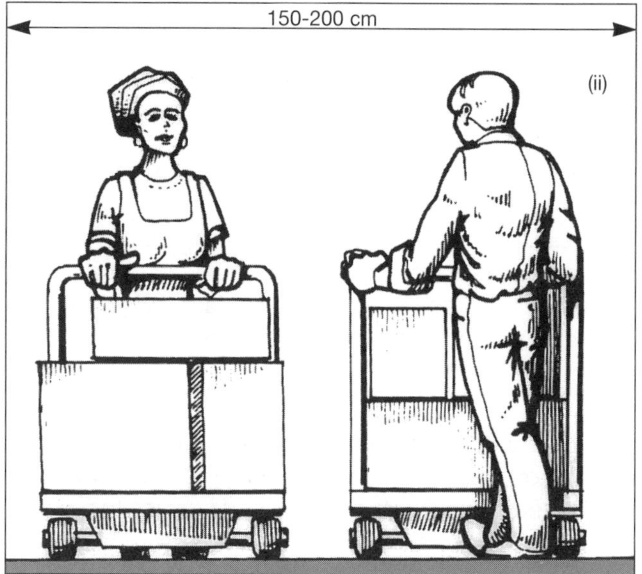

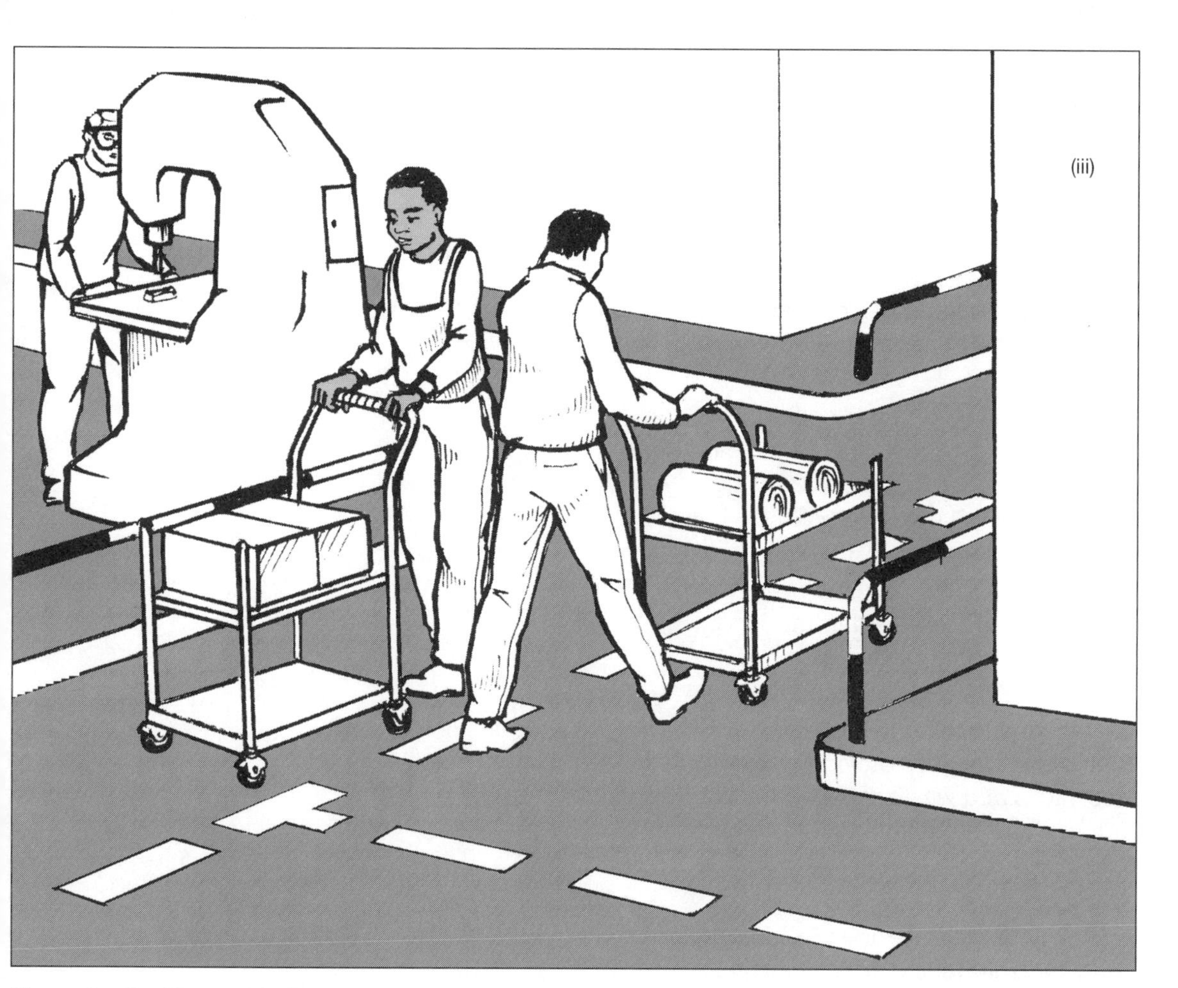

Figure 2. (i), (ii) and (iii) Keep aisles and corridors wide enough to allow two-way transport

CHECKPOINT 3

Make the surface of transport routes even, not slippery and without obstacles.

WHY

Transport within the enterprise is an important part of day-to-day work. The smooth flow of transport from the storage area to worksites, and between workstations, is a prerequisite for a productive workplace.

Carrying loads on an uneven or slippery floor is a common cause of accidents. Such accidents are eliminated by arranging good transport routes.

Products may fall when the workers stumble or hit obstacles, thus causing loss of production or increased cost of repair.

The use of carts and wheeled racks is considerably easier if the surface is even and free from obstacles.

HOW

1. Remove sudden height differences or other stumbling hazards in the transport routes.

2. Make it a routine to remove or avoid spilled water, oil or other slippery substances (by cleaning, laying easy-to-clean floor surfaces or using absorbent materials). Use tight, covered transport containers to avoid spills.

3. If uneven spots cannot be removed immediately, use ramps, fill-ins or loading platforms.

4. Make it a rule to place nothing in the aisles or corridors. This is best done by providing good storage places, racks and waste receptacles in sufficient numbers, and by defining and marking transport routes.

5. Promote the use of transport devices including carts, mobile racks, trolleys and small vehicles. Large wheels are preferable to small ones except for short-distance transport on hard, even surfaces.

SOME MORE HINTS

- Transport surfaces can be covered or painted with high-friction coatings which reduce the risk of slipping but do not influence rolling resistance of carts and trucks.

- Bright painting of transport surfaces makes it easy to identify slipping risks. Adequate lighting helps to identify unevenness.

- Unsteady or low-friction footwear may cause slipping, even on good surfaces. If the slipping risk is considerable, provide workers with adequate shoes.

POINTS TO REMEMBER

Cleaning of the transport routes is a low-cost solution to important problems. Make cleaning a daily routine.

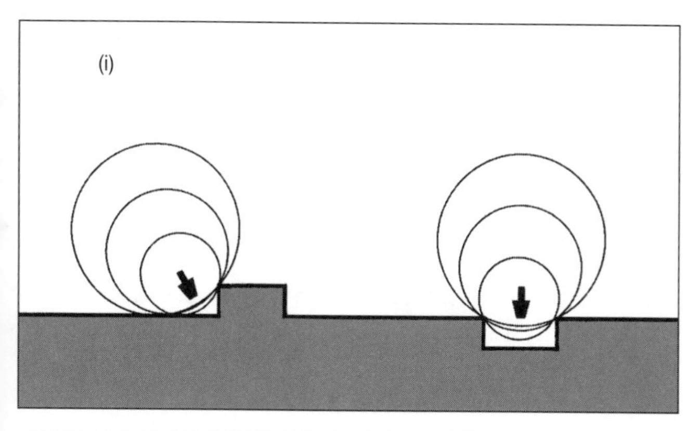

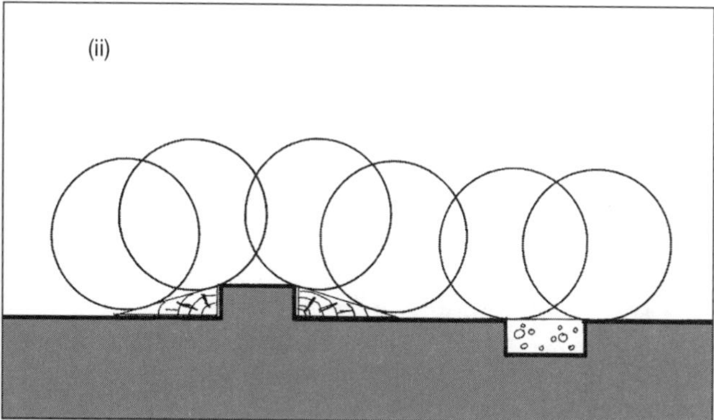

Figure 3a. (i) Remove sudden height differences wherever possible. Larger wheels are generally better than smaller ones as they can more easily overcome any obstacles or hollows that happen to remain. (ii) Fill in or bridge sunken places. If height differences remain, provide gradually inclined covers so as to avoid stumbling or wheel obstacles

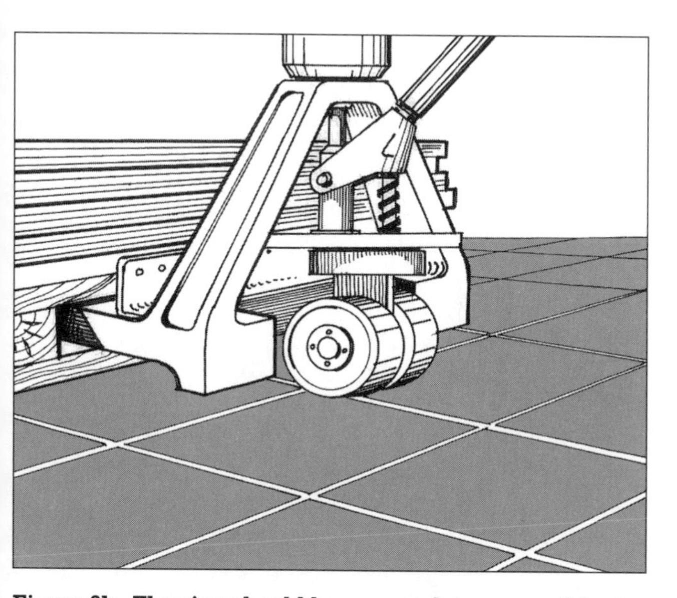

Figure 3b. Flooring should be as complete as possible, to eliminate stumbling obstacles or sunken places

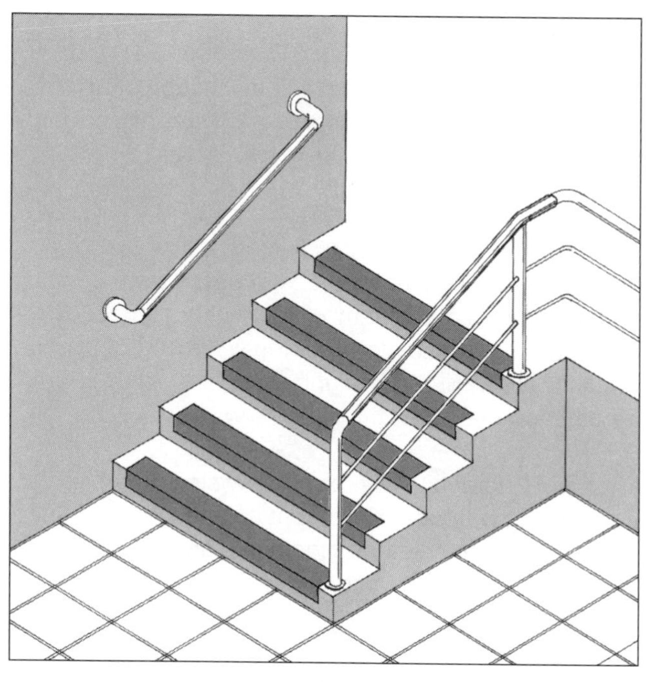

Figure 3c. Make arrangements to prevent slipping on slopes or stairs. High-friction materials placed at the edges of stairs can help

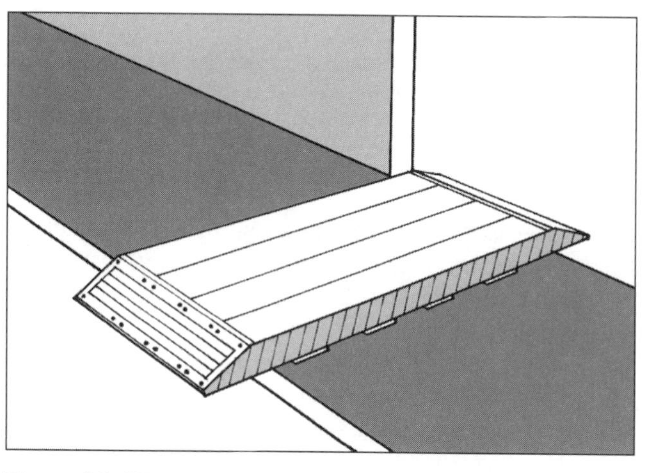

Figure 3d. Where bridging is needed, make sure that the surface is not slippery and allows the passage of wheels

CHECKPOINT 4

Provide ramps with a small inclination of up to 5 to 8 per cent instead of small stairways or sudden height differences within the workplace.

WHY

Sudden height differences in passageways within the workplace hamper smooth transport of materials and can cause accidents. Instead of posting "MIND THE STEPS" here and there, provide ramps eliminating the hazard.

Short stairways with only a few steps may seem easy to climb up and down, but can cause stumbling or falling that may lead to accidents and damage to products. It is worth while considering the use of ramps instead of stairs.

Ramps with a small inclination make it possible to use push-carts and wheeled racks. This greatly facilitates the transport of work items.

HOW

1. Where there are small sudden height differences or stairways with just a few steps, replace them by a ramp with a small inclination of up to 5 to 8 per cent.

2. Make sure that there are no stumbling obstacles on entering or leaving the ramps provided. Also ensure that the surface of the ramps is adequate and non-slippery.

3. If there is a danger of falling from the side of the ramp, provide fences or hand-rails.

4. Encourage the use of carts and mobile racks instead of manual carrying of materials and semi-products. Ramps are perfectly suited for these.

SOME MORE HINTS

- Avoid slippery ramp surfaces. Make sure that the ramp surfaces do not get wet.

- When carts or mobile racks are used, provide firm grips or handles to ensure easy and safe transport on the ramps.

- Check the workplace layout and means of transport so as to reduce the frequency of transport, especially when transport is necessary between workplaces with height differences.

POINTS TO REMEMBER

Ramps can prevent stumbling and facilitate transport operations. They lead to fewer and safer transport trips through the use of pushcarts or mobile racks.

Figure 4. Provide ramps instead of stairways

CHECKPOINT 5

Improve the layout of the work area so that the need to move materials is minimized.

WHY

Often machines and workstations are installed one after another as production expands, and their existing positions are not necessarily suitable for easy and efficient movement of materials. This can be improved by changing their layout.

Time needed to perform a task can be greatly reduced by reducing the movement of materials. This lessens workers' fatigue, allowing more efficient working.

This is also beneficial for preventing accidents caused by moving materials.

HOW

1. Discuss with workers how the frequency and the distance of moving materials can be reduced by changing the layout of machines and workstations. There should be a better way of moving materials within work areas and between different work areas.

2. Arrange the locations of a series of several workstations so that the work items coming from the previous workstation can go directly to the next work area.

3. Arrange the locations of different work areas according to the sequence of work done so that the work items coming from one work area can be utilized by the next work area without moving them over a long distance.

4. Combine operations whenever possible in order to reduce the need for moving materials between operations.

SOME MORE HINTS

- Use pallets or a stock of work items so that items coming from one workstation can be moved easily to the next workstation or work area.

- Ensure that transport routes are clear when rearranging the layout of the work area.

- A flexible work area layout that can be adapted to changes in work flow (for example, because of product changes or in order to produce several different products) is a productive layout.

POINTS TO REMEMBER

Minimizing the need to move materials by improving the layout of the work area is the surest way to reduce time and effort, and increase productivity.

Figure 5a. Provide stock shelves or racks so that work items coming from one workstation can go directly to the next one

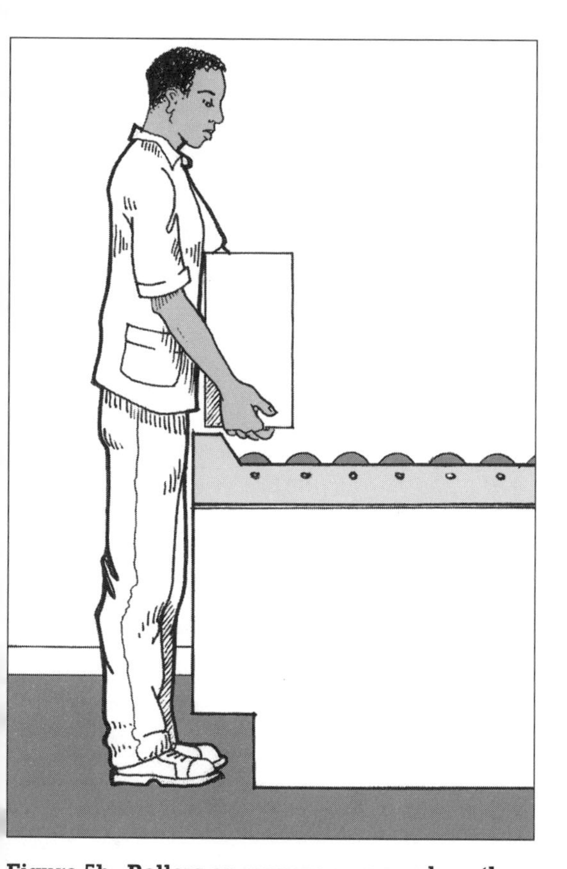

Figure 5b. Rollers or conveyors can reduce the distance of moving materials manually. The height should be appropriate for handling the work item without bending the upper body. Make sure that there is enough space for getting close to the rollers or conveyors and for the feet

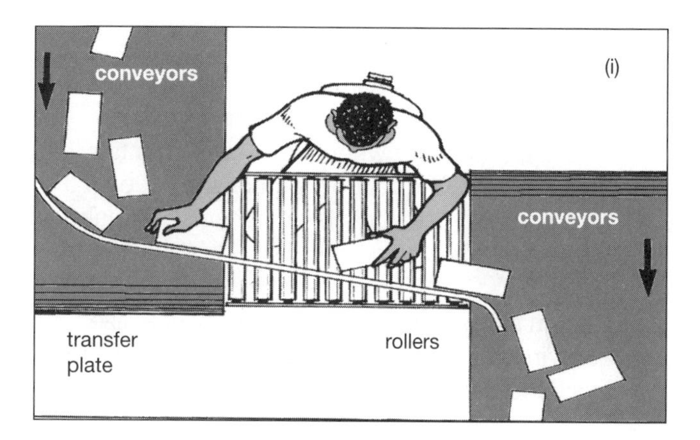

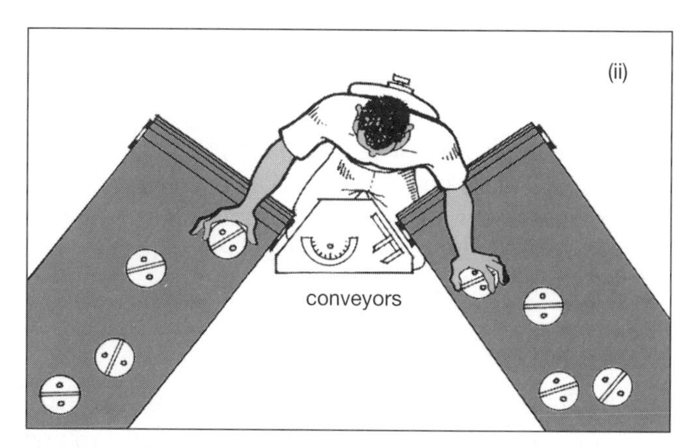

Figure 5c. (i) and (ii) Provide the layout which allows the worker to move objects from one conveyor to the next while keeping a natural posture. The use of a transfer plate and rollers may help ease the moving of objects

CHECKPOINT 6

Use carts, hand-trucks and other wheeled devices, or rollers, when moving materials.

WHY

Moving many materials not only consumes a lot of effort, but often leads to accidents which damage materials and may injure workers. All this is avoided by using "wheels".

By using carts and other mobile devices the number of trips can be significantly reduced. This means improved efficiency and safety.

Rollers placed one after another along a materials movement line greatly ease the movement of materials because only the pushing and pulling of rollers is needed instead of carrying.

HOW

1. Check movements of materials between the storage and work areas, and between work-stations, especially when these movements of materials are frequent or require a lot of effort. Consider the use of carts or "wheels" to make these movements easier.

2. Design simple pushcarts of appropriate size for carrying materials. Construct such carts using available parts and skills.

3. Provide a line of rollers on which materials can be pushed easily to the next workstation. A 2 m long roller line can be a great help.

4. Use pallets, bins or containers that can be loaded easily onto a pushcart or pushed through on rollers. Design special ones for different products so that the products are protected from damage, and are easy to count and inspect.

SOME MORE HINTS

- It is important to have clear transport routes free from obstacles at all times. Clear transport routes are essential for moving around with a cart.

- Materials can be moved between workstations by conveyers, rollers, gravity chutes, suspended cranes, mobile hoists and other devices. There are many ways to construct such systems at low cost.

- A long, rectangular, mobile frame on which rollers are placed one after another could also be used for loading and unloading trucks.

- Choose wheels of a larger diameter, especially when moving materials a long distance or on uneven surfaces.

- If possible, adopt rubber wheels or castors to reduce noise.

POINTS TO REMEMBER

Reduce the number of trips between work-stations and between storage and work areas by using wheeled transport such as pushcarts, or roller lines.

Figure 6a. (i) A heavy-duty sack truck and (ii) a low-lift pallet trolley are reliable, safe and easy to operate. They provide means of carrying heavy loads a short distance with minimum elevation

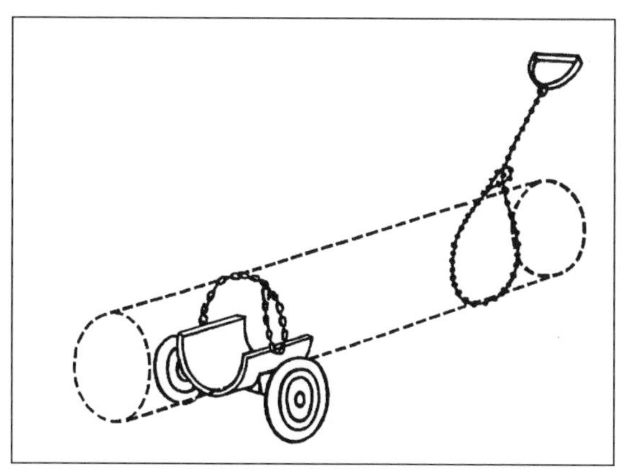

Figure 6c. This small cart enables one worker to move heavy metal bars

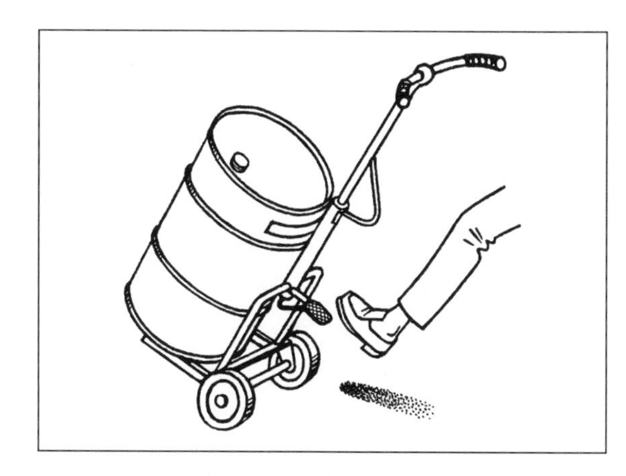

Figure 6d. This barrel-handling device not only makes work much easier but also helps avoid damage

Figure 6b. A passive conveyor line for moving heavy motor castings at working height

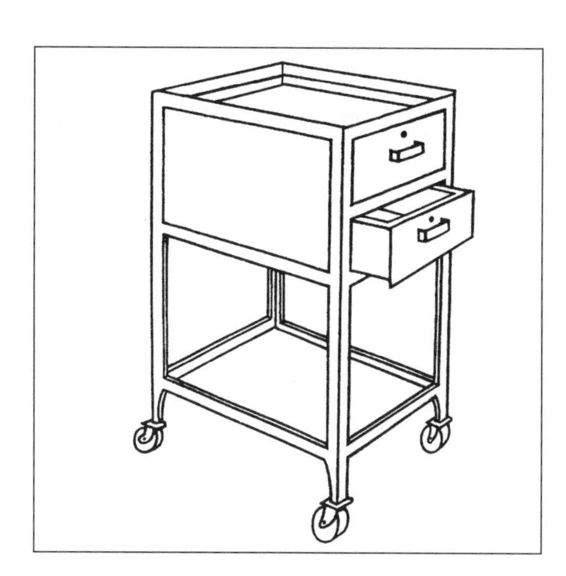

Figure 6e. An easy-to-move tool cart provides orderly storage and protection of tools and instruments

CHECKPOINT 7

Use mobile storage racks to avoid unnecessary loading and unloading.

WHY

There are usually a large number of items that need to be carried to other workstations or to storage areas. If these items are put on mobile racks and the racks are then moved to the next place, many unnecessary trips can be avoided.

Carrying work items together on mobile racks means fewer materials-handling operations (such as loading and unloading). This helps to reduce damage to work items, minimize accidents and use workers' energy for productive operations.

Using mobile racks also means easy inventory control and more efficient housekeeping.

HOW

1. Design or purchase racks that have wheels and can move a number of items at a time. Choose racks which are easy to load and unload.

2. Arrange the workplace layout so that it allows the smooth movement of wheeled racks between workstations, and between storage and work areas. If necessary, redefine the transport routes.

3. When many small items are to be moved, provide adequate space for each item so that all items can be placed neatly on the mobile rack.

4. Consider the use of pallets, containers, trays or bins that can be placed on a mobile rack or a pushcart.

5. Fit wheels to existing stands, racks or workbenches in order to make them mobile and thus avoid unnecessary loading and unloading operations.

SOME MORE HINTS

- If designing effective mobile racks seems complicated, a good first step may be to design pallets or trays to move several items at the same time. This experience will make it easier to design a rack that is both easy to handle and efficient.

- When many similar racks are used, standardize the racks. Similarly, when many pallets or containers are used for work items, standardize them so that they can be easily placed on a mobile rack or cart. It is even better if these pallets or containers can be stacked.

- Maintenance of wheels or rollers is very important because it makes pushing and pulling easier.

- It is worth while to invest in designing special-purpose mobile racks for particular work items, even though you might feel that this requires money and effort. Such racks are extremely helpful for improving productivity. They enable many items to be placed on the rack by easy handling operations and moved conveniently to other worksites.

POINTS TO REMEMBER

Mobile storage racks are an ideal answer to reducing handling operations and transport time. Benefit from mobile racks.

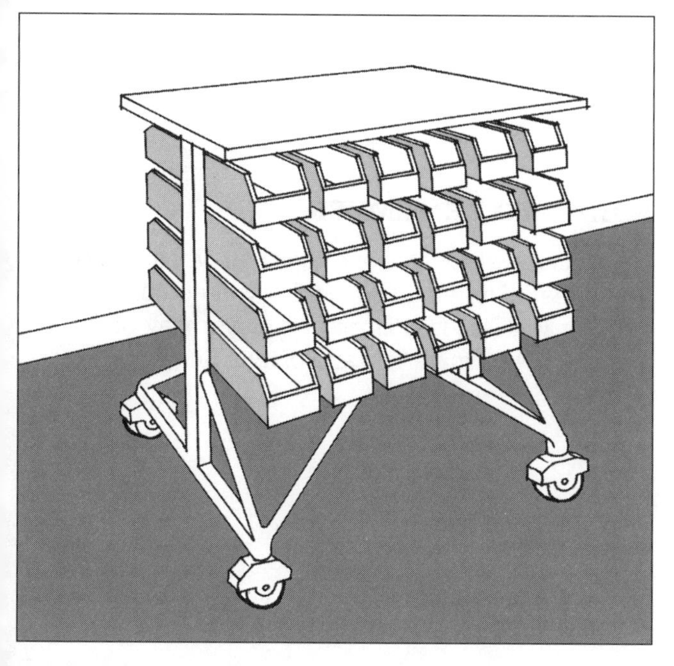

Figure 7a. A mobile bin cart helps to ensure smooth work flow in assembly shops where numerous operations are performed at each workstation

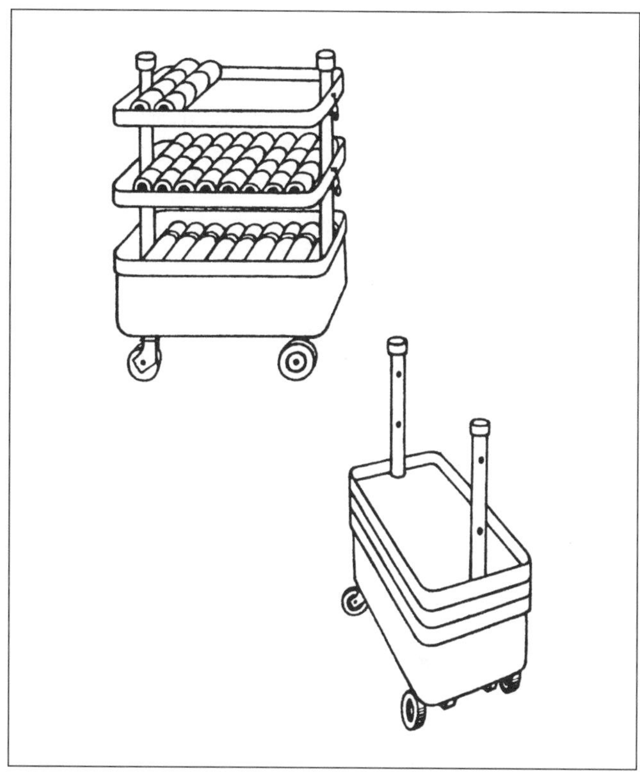

Figure 7b. A tool trolley with adjustable shelves occupies little space, but contributes much to improving the efficiency of motor mechanics and machine tool repair workers

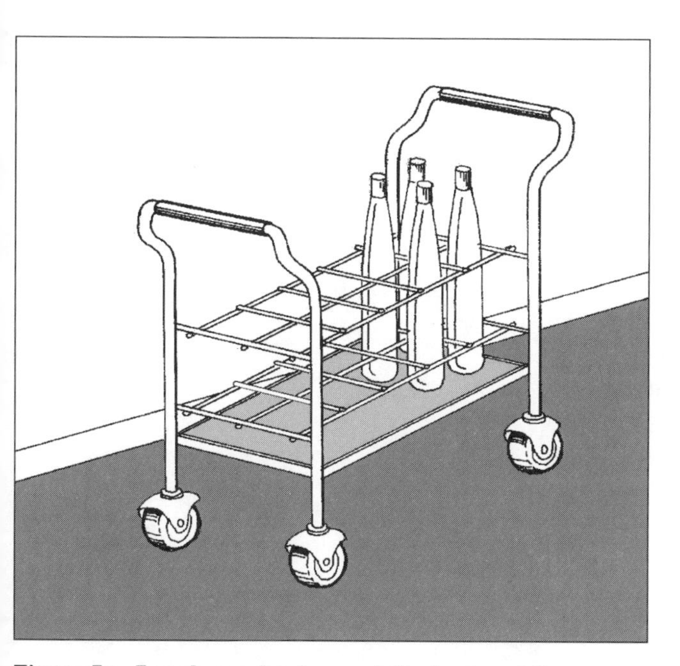

Figure 7c. A rack on wheels specially designed for storage and handling of motorcycle silencers

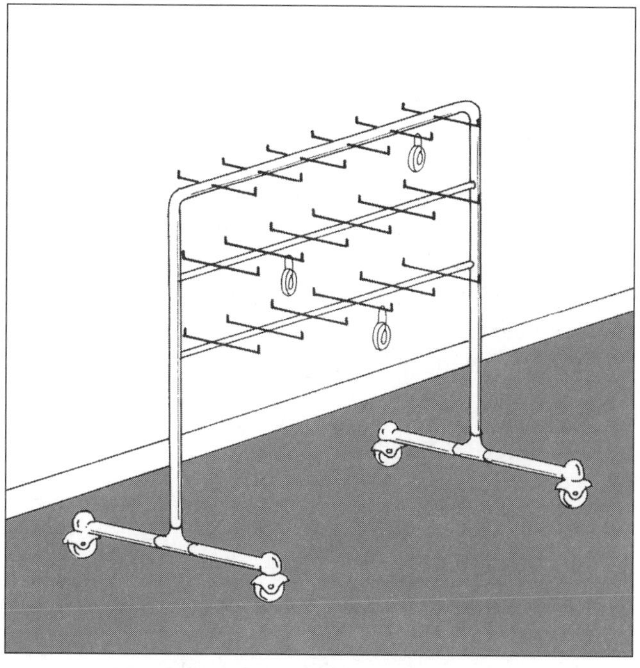

Figure 7d. A flat, two-sided movable rack, a real "space saver" for a small factory with narrow passages, can be successfully used for many types of workpiece

CHECKPOINT 8

Use multi-level shelves or racks near the work area in order to minimize manual transport of materials.

WHY

Placing materials near the workstation, so that they are easy of access and at an appropriate height, can save time and energy spent in picking them up.

Multi-level shelves and racks allow a better use of space and help to keep things in good order when space available near the work area is limited.

Shelves and racks with space specified for each individual item are excellent for safe storage of materials and semi-finished products, especially fragile ones; this reduces the danger of accidents and fires, and the possibility of damage.

HOW

1. Provide multi-level, open-fronted shelves or racks for various specific items.

2. Make full use of wall space by fitting multi-level shelves or racks to the wall near the work area.

3. Wherever possible, make racks movable by fitting wheels to them.

4. Provide a different, specially arranged place for each kind of material or part so that access to them, as well as their stocking and transport, is easy; use labels or other means to indicate each specific place. Avoid levels that are too high or too low as they are then difficult to reach.

SOME MORE HINTS

- Use lightweight containers and bins for storage of small parts. Front-opening containers and bins make the materials and parts inside easy to see and to grasp.

- Use pallets or trays with an individual space for each specific item for easy storage, easy access and easy stocking.

- Store heavy or awkward items at waist height or in a way appropriate for the next stage of transport; store light and infrequently used items at knee or shoulder level.

POINTS TO REMEMBER

Multi-level shelves and racks on wheels save a great deal of time and space. They keep things in good order. This is a simple yet clever way of reducing damage to materials and of avoiding accidents.

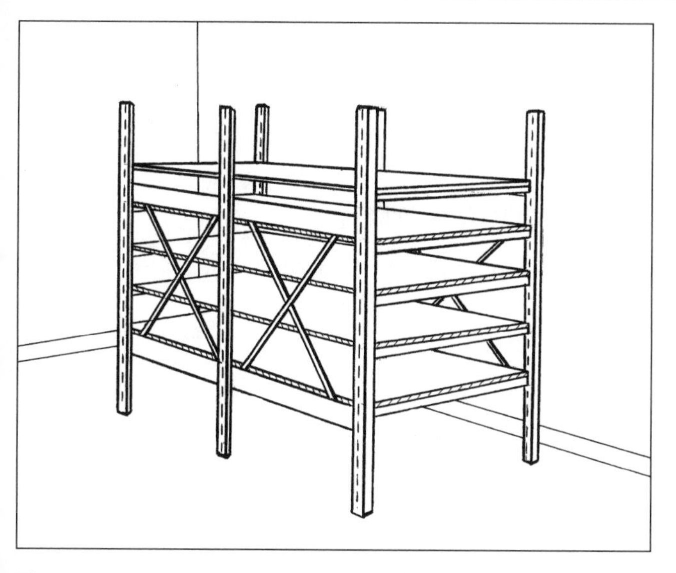

Figure 8a. A multi-level horizontal storage rack for metal sheets or plywood. Remember to keep everything dry, otherwise water tends to spread between the sheets and damage them

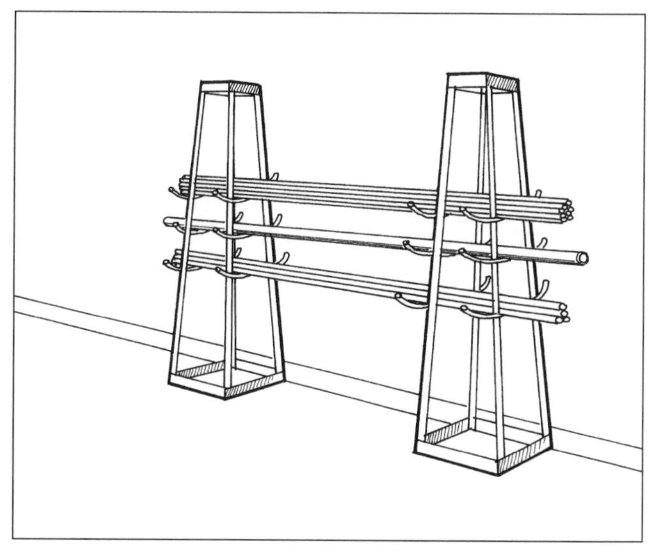

Figure 8b. A horizontal bar rack. This free-standing unit may be used singly to store short pieces, or two of these racks may be placed in line to store long pieces

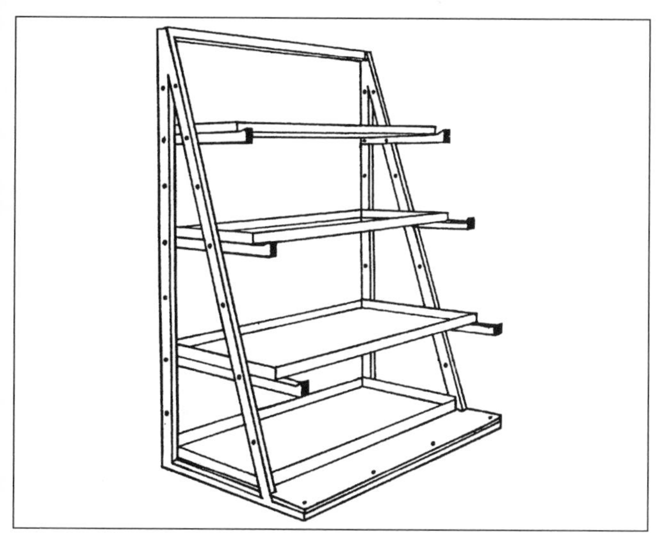

Figure 8c. A vertical rack. Metal rods and bars of different profile can be stored efficiently in a limited area or near the job. Tray-type shelves provide room for small pieces

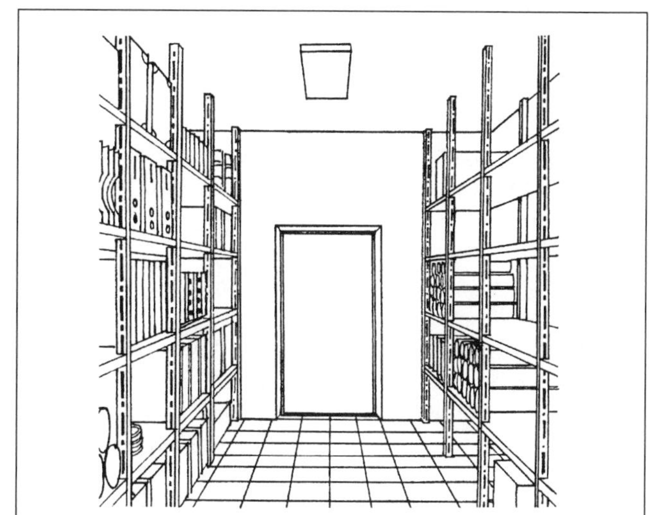

Figure 8d. Shelving designed to use wall space fully

Figure 8e. A shop floor after removal of all unnecessary items. All tools and parts are stored on shelves and racks

CHECKPOINT 9

Use mechanical devices for lifting, lowering and moving heavy materials.

WHY

Manual lifting, lowering and moving of heavy materials and work items are some of the major causes of accidents and back injuries associated with materials handling. The best way to prevent these accidents and injuries is to eliminate the manual work by using mechanical devices.

Lifting and carrying heavy objects manually requires skill and takes much time. With mechanical means, these tasks are done more efficiently and quickly.

The introduction of mechanical devices for handling heavy materials greatly helps to organize the work flow.

HOW

1. Install floor-based lifting devices which use the minimum elevation necessary. Examples are gantries, hydraulic lifting devices, lift-tables, hydraulic floor cranes, lever or chain hoists, electric hoists, or conveyers.

2. Overhead cranes and overhead hoists can be used if the workplace structure permits them. However, consider that these overhead devices bring hazards to the workplace that may result in serious accidents. If possible, floor-based lifting devices are better as they can be used with much less elevation of materials.

3. Use only lifting machinery which has been tested by the maker or some other competent person, and for which a certificate specifying the safe working load has been obtained.

4. Make sure that the maximum safe working load is plainly marked and that it is observed.

5. Make sure that qualified persons regularly inspect and maintain the lifting machines, chains, ropes and other lifting tackle.

SOME MORE HINTS

- Manual lifting of heavy loads should be considered as a last resort in special cases when the application of mechanical means is not feasible.

- Lifting heavy loads is usually combined with transport. Organize the lifting so that the next step of transport is easier. Lifting to the working level on a mobile lift-table is a good example.

POINTS TO REMEMBER

Use mechanical lifting devices for the minimum elevation necessary for safety and efficiency.

Figure 9a. A portable gantry is reliable, safe and easy to operate for carrying a heavy load a short distance with minimum elevation

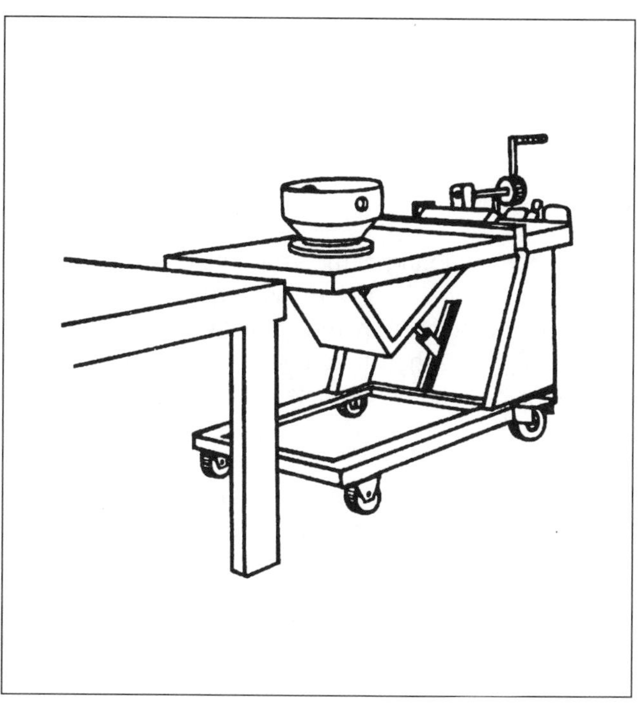

Figure 9b. A manually powered device for lifting heavy castings to working level

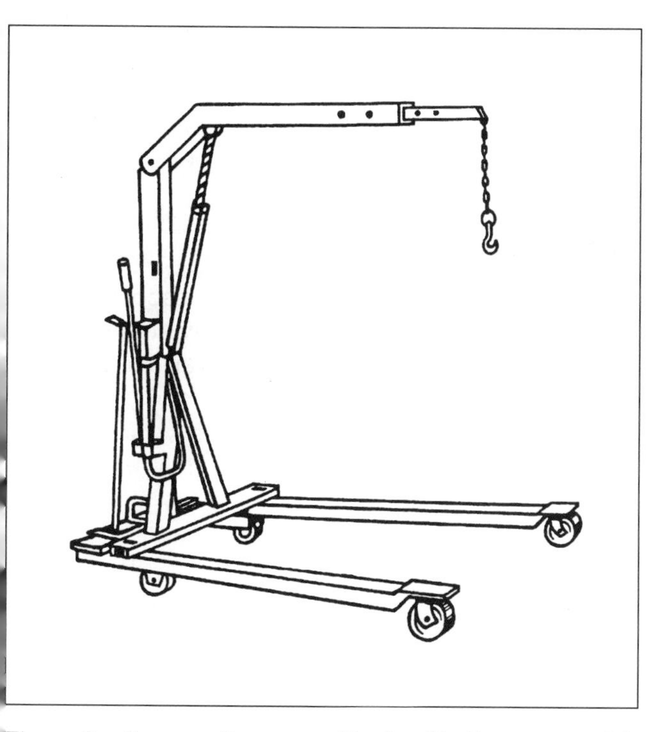

Figure 9c. A manually powered hydraulic floor crane with a telescopic boom

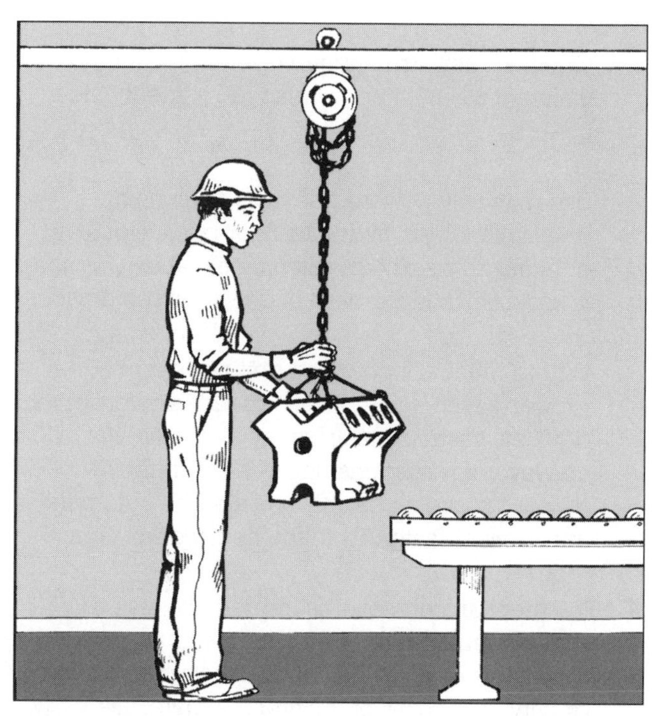

Figure 9d. Make sure that the maximum safe working load is plainly marked

CHECKPOINT 10

Reduce manual handling of materials by using conveyers, hoists and other mechanical means of transport.

WHY

Manual handling of materials, including stocking, loading and unloading, does not add any value or profit. By replacing it by mechanical means, the workers can use their time for productive work. This is true for both heavy objects, and light and small objects.

Repetitive manual handling of materials involves bad working postures and frequent awkward movements. This may cause muscle and joint problems, resulting in poor productivity. Using equipment instead of manual handling greatly reduces fatigue and the risk of injury.

Mechanical transport of materials improves productivity a great deal and thus offers the chance of organizing a better work flow.

HOW

1. Check materials-handling operations to see which of them can be replaced by mechanical means.

2. Use manually powered devices, such as hydraulic lift-tables, hydraulic floor cranes, or lever or chain hoists. The maintenance of these devices is easier than in the case of power-driven devices.

3. Where manually powered devices cannot deal properly with heavy materials, use power-driven devices, such as electrical or hydraulic lifts, conveyers or suspension lines. Often these devices make it possible to organize automatic transport of materials to the next workstation.

4. If a mechanical mover is impractical, use a gravity chute for light materials and an inclined roller conveyer for heavy materials. The force of gravity takes care of moving the materials.

5. Train workers in safe procedures for using the mechanical means of transport. Also make sure that there is enough space for safe operations.

6. Make sure that the dangers presented by new mechanical devices are evaluated properly and that adequate countermeasures are taken.

SOME MORE HINTS

- Use mechanical devices that can be easily operated by different workers for different handling tasks. This facilitates use of the devices.

- Install the mechanical devices so that manual work prior or subsequent to mechanical handling is easy, e.g. so that the worker need not lift or lower the materials any further.

- Use a pushcart or trolley whereby the materials can be brought to the stocking or unloading point at the right height. If appropriate, consider the use of a special stand or platform of the right size and correct height placed near each machine. Materials carried to the machine can be easily stocked on such a stand or platform.

- Transport and supply of toxic or dusty materials require special attention. The use of closed containers and the enclosure of areas where workers eat meals or snacks should be considered.

- Learn from good examples already in use on similar machines. There should be many simple and practical ideas.

POINTS TO REMEMBER

By using mechanical means of transport, the worker's productive hands and energy are released from handling materials and made available for more profitable and safer tasks.

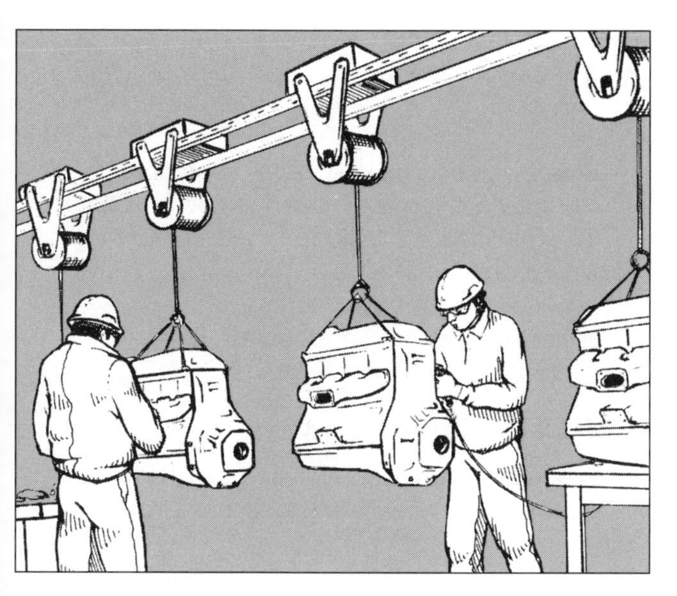

Figure 10a. Mechanical transport of materials can both eliminate the manual work and improve work height and working posture

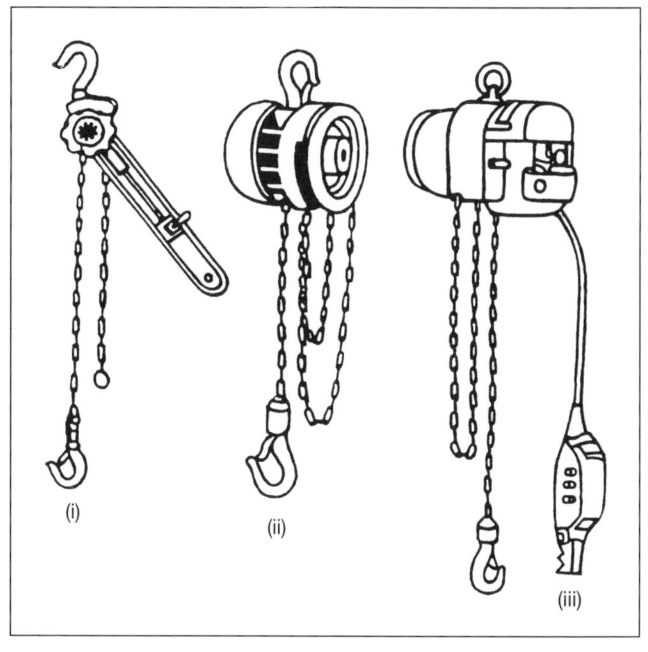

Figure 10b. (i) A lever hoist is simple to operate and extremely versatile. (ii) A chain hoist with a self-activating load brake. (iii) An electric chain hoist with butterfly control switch for efficient handling of lighter loads

Figure 10c. Reduce manual handling of materials by using conveyors

CHECKPOINT 11

Instead of carrying heavy weights, divide them into smaller lightweight packages, containers or trays.

WHY

Carrying heavy items is strenuous and dangerous. This makes the job very unpopular. Divide heavy items into smaller objects whenever possible.

The fatigue from carrying lightweight packages is much less than the fatigue from carrying heavy weights. The worker's productivity is improved by carrying lighter packages.

The risk of back injuries is also greatly reduced by using light instead of heavy packages.

HOW

1. Check all manually lifted or carried weights to see the possibility of dividing them into smaller weights.

2. Divide heavy loads into lighter packages, containers or trays, considering the maximum weight with which the worker is comfortable. For example, two packages of 10 kg each are far better than one package of 20 kg.

3. When loads are divided into smaller packages, this may mean increased movements and more trips for carrying the same total amount. Therefore, be sure that packages are not too small, and that effective means of moving or carrying these smaller packages are used.

4. Use carts, trolleys, mobile racks or hand-trucks for carrying many packages at one time. This is in line with the idea of dividing heavy packages into smaller ones, as the total amount carried by carts, etc., does not decrease overall, although loading and unloading are easier and faster.

SOME MORE HINTS

- Organize the use of transport devices by studying the situation jointly with workers in order to avoid manual carrying whenever possible.

- Ensure that packages have good hand grips, so that the load can be kept near the waist while carrying.

- Smaller objects make it easier to organize the workplace in terms of materials flow and storage. They also reduce accidents such as dropping or stumbling.

- Make sure that transport routes are even, not slippery and without obstacles.

POINTS TO REMEMBER

A lighter weight is a safer weight. Divide heavy weights into lightweight packages to improve safety and efficiency at work.

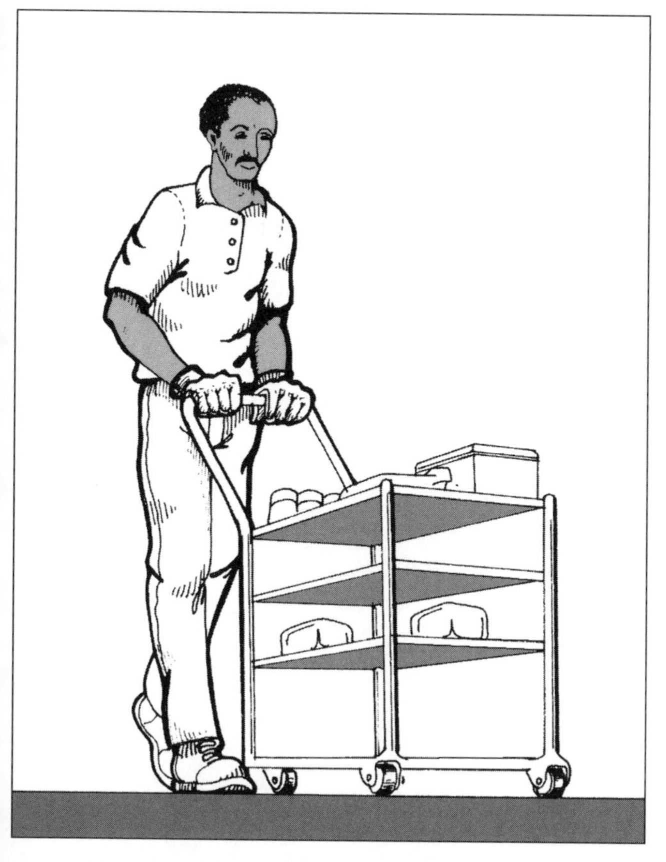

Figure 11a. Divide heavy packages into smaller lightweight ones

Figure 11b. Divide heavy items into smaller objects whenever possible

CHECKPOINT 12

Provide handholds, grips or good holding points for all packages and containers.

WHY

Carrying loads is much easier and quicker if they can be grasped easily and firmly.

With good grips there is less chance of dropping the loads, and thus damage to materials is prevented. Good hand grips also provide a clear forward view.

Good handholds make it possible to reduce fatigue, since there is less bending of the body and less muscle power required to hold the load.

HOW

1. Cut out handholds in boxes, trays and containers so that the boxes, etc., can be comfortably carried by the hands.

2. Use packages that have a grip or good holding points for carrying.

3. Ask suppliers and subcontractors to deliver goods in boxes or containers with handholds or grips.

4. Locate handholds so that it is possible to carry the load in front of the body.

5. When a load is carried by means of one handhold or grip, locate it so that the centre of gravity of the load is close to the worker's body.

SOME MORE HINTS

- Make it a rule to order boxes, trays and containers that have handholds or grips.

- Consider putting the handhold at an angle by which the box or container can be carried with the wrist in a comfortable, straight position.

- Packaging should be designed to simplify manual handling (for example, in lightweight packages) and to provide grips or good holding points. Slippery package surfaces should always be avoided.

- Be aware that the worker may be wearing gloves. Grips or handholds must be easily grasped by gloved hands.

POINTS TO REMEMBER

Fitting handholds on boxes and containers is a very simple measure to improve materials handling.

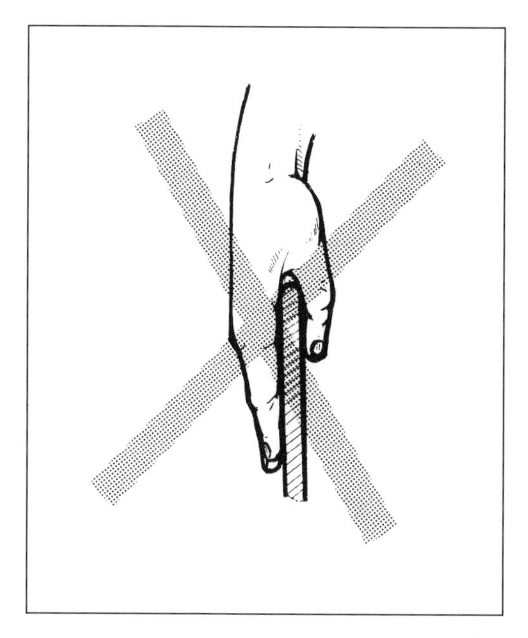

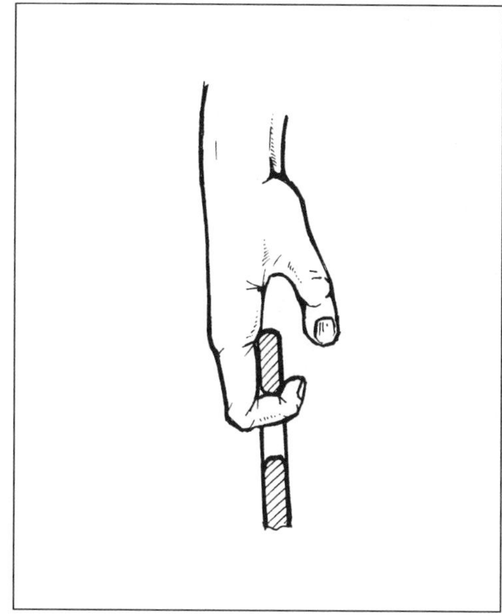

Figure 12a. Handholds should be cut out so as to allow gripping of containers by bent fingers. This can greatly reduce the force needed to hold the containers

Figure 12b. Cut-out handholds are very useful. Locate these handholds so that the box or container can be carried in front of the body

CHECKPOINT 13

Eliminate or minimize height differences when materials are moved manually.

WHY

Materials handling is an important and common element of any enterprise's activities. If it is done effectively, it can ensure a smooth flow of work. However, materials handling takes time and energy, often resulting in damage, delays and even accidents. One of the causes of bottlenecks is lifting and lowering movements. By minimizing these movements, problems related to materials handling become far fewer.

By avoiding lifting and lowering movements, you can reduce workers' fatigue and material damage, as well as increase the efficiency of materials movement.

Manual lifting is one of the most strenuous activities at work and an important cause of accidents and back injuries. By minimizing it, you can also reduce the risk of injury and absenteeism.

HOW

1. When materials are moved from one workstation to another, move these materials at the working height. For example, move between work surfaces of the same level.

2. If large items are placed on the floor, use a yoke, sack, hand-truck or low-level pallet trolley to carry them with minimum elevation.

3. Use transport systems whereby materials can be moved without changing the height. Examples are a passive conveyer line (using rollers placed at the same level), a mobile work-stand or trolley that is of the same height as the work-tables, or suspension of materials that move at the same level.

4. Match the height of the vehicle bed to that of the loading area, so that loading and unloading can be done with minimum height differences.

SOME MORE HINTS

- Use work-stands or platforms for placing materials so as to reduce height differences in moving these materials.

- Build special materials-handling devices adapted to your work items that allow a minimum of elevation of the load. Examples are cylinder-carrying hand-trucks, barrel carrying pushcarts or a mobile suspension for heavy items.

- Use mechanical lifting devices whereby the height for manual materials movement can be adjusted, such as lift-trucks or adjustable conveyers.

- When designing new work areas, eliminate height differences of working surfaces.

POINTS TO REMEMBER

Move materials at working height. Use mechanical devices to raise or lower materials to this working height.

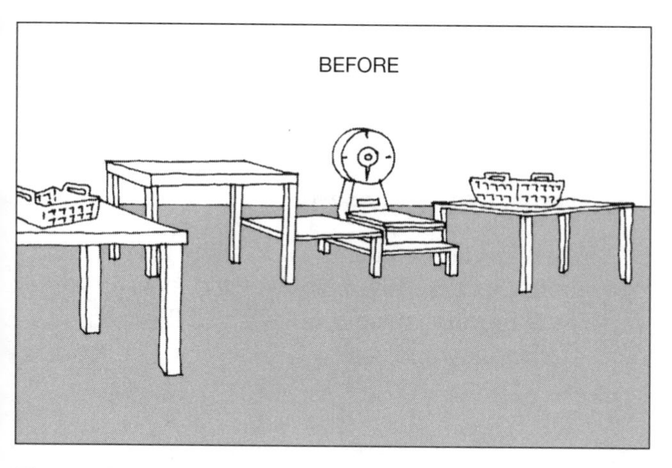

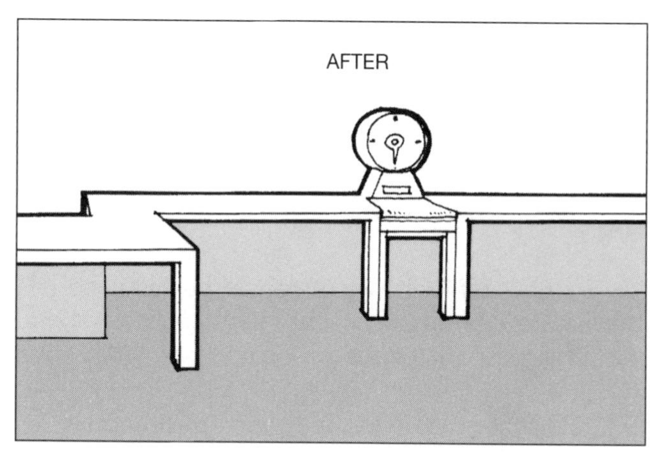

Figure 13a. Eliminate height differences of work surfaces

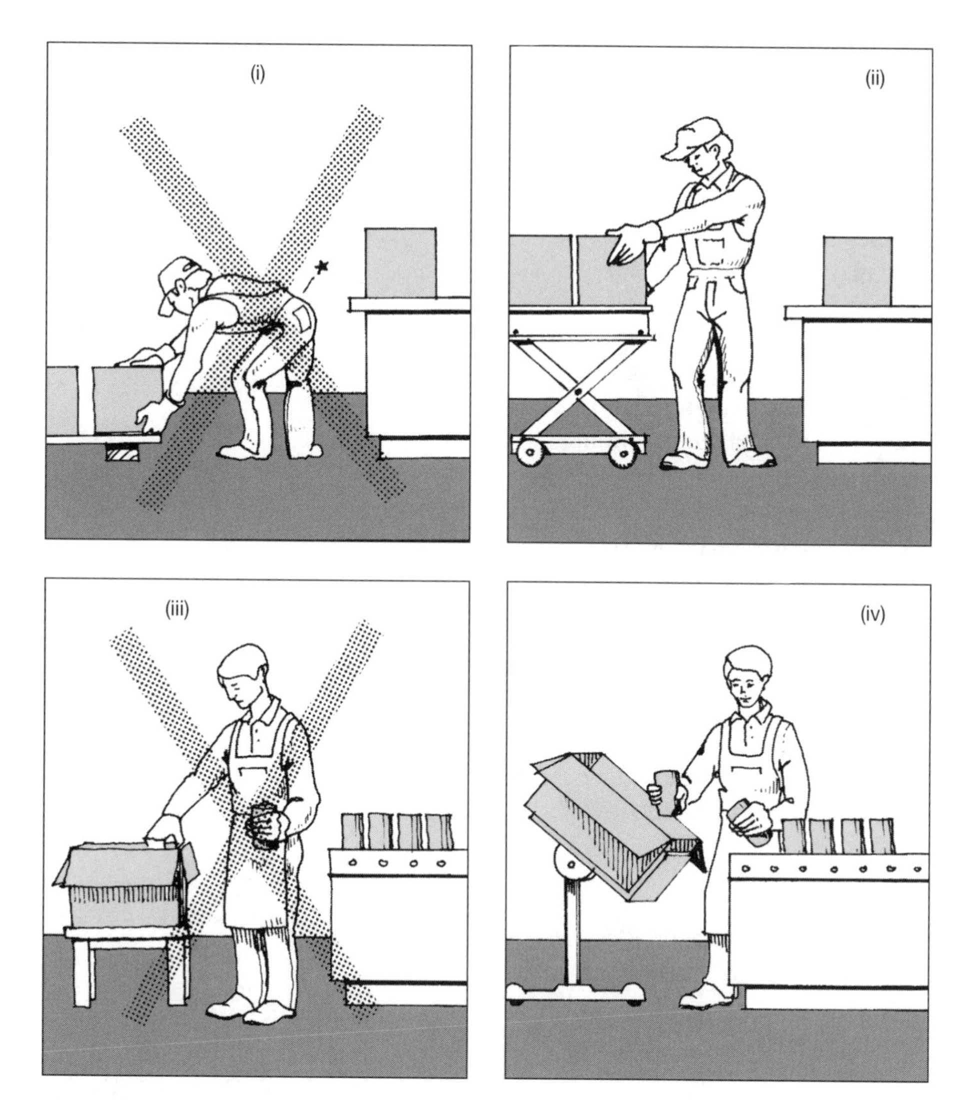

Figure 13b. (i) and (ii) Eliminate or minimize height differences. (iii) and (iv) Minimize lifting and lowering movements

CHECKPOINT 14

Feed and remove heavy materials horizontally by pushing and pulling them instead of raising and lowering them.

WHY

Particularly for heavy materials, pushing and pulling are less strenuous and safer than lifting and lowering these materials.

Horizontal moving of heavy materials is more efficient and allows better control of work, because the work requires less force and the worker does not need to move the weight of the body.

Pushing and pulling at appropriate height, rather than lifting, helps to prevent back injuries.

HOW

1. Use transport devices, such as conveyers or pushcarts, to bring heavy materials to the place where they can be fed into the machine at the right height.

2. If mechanical devices cannot be used to feed heavy materials into the machine, use a roller-conveyer which makes the horizontal movement of the materials easier.

3. Ensure that there is enough space for the feeding and unloading positions at the machine so that gliding or sliding of heavy, repeatedly handled objects can be done easily. Note that an even and non-slippery floor surface is important for efficient and safe work.

4. Use simple lifting devices or mobile lift-tables to move the load to the feeding or workbench level.

SOME MORE HINTS

- Avoid handling of heavy materials in confined spaces as it limits moving efficiency and may cause bad posture and accidents.

- Pushing or pulling is more efficient when it is done forwards and backwards rather than sideways in relation to the body.

- When moving heavy work items from one workstation to the next, keep them at working height so that there is no need for raising or lowering movements.

POINTS TO REMEMBER

When feeding and unloading stock for machine operations or assembly, move heavy materials horizontally, rather than raising or lowering them, for higher efficiency and safety.

Figure 14. Push and pull heavy materials instead of raising and lowering them

CHECKPOINT 15

Eliminate tasks that require bending or twisting while handling materials.

WHY

Bending or twisting of the body is an unstable movement. The worker spends more time and becomes more fatigued than when doing a similar amount of work without bending or twisting.

Bending and twisting of the body is one of the major sources of back injuries, and neck and shoulder disorders.

HOW

1. Change the positions of materials or semi-products so that the handling work is done in front of the worker without bending the body.

2. Improve the working space for doing the handling work so that the worker can adopt stable foot positions without bending or twisting.

3. Use mechanical means to bring the work items to the front of the worker. The worker should be able to remove the work item and to replace the finished item without being forced into an awkward posture.

4. Change the working height (e.g. by changing the height of the work-table or feeding point) so that the worker can handle the work item without bending the body.

SOME MORE HINTS

- Avoid manual handling and carrying of heavy objects whenever possible. Bending or twisting while dealing with heavy objects is particularly harmful.

- Sometimes standing workers bend their body because they cannot come close enough to the work item owing to a lack of knee or foot clearance. Make sure that the workers have enough clearance for their knees and feet.

- Avoid a combination of carrying while performing other tasks at the same time, as this is often the reason why the worker bends or twists the body. Rearrange the work so that the carrying task is the only task performed at that time.

POINTS TO REMEMBER

Back injury resulting from bending or twisting while handling a heavy load can cost you a great deal, as you may lose a productive skilled worker for quite a long period.

Figure 15a. (i) and (ii) Minimize the distance between the worker and the work item

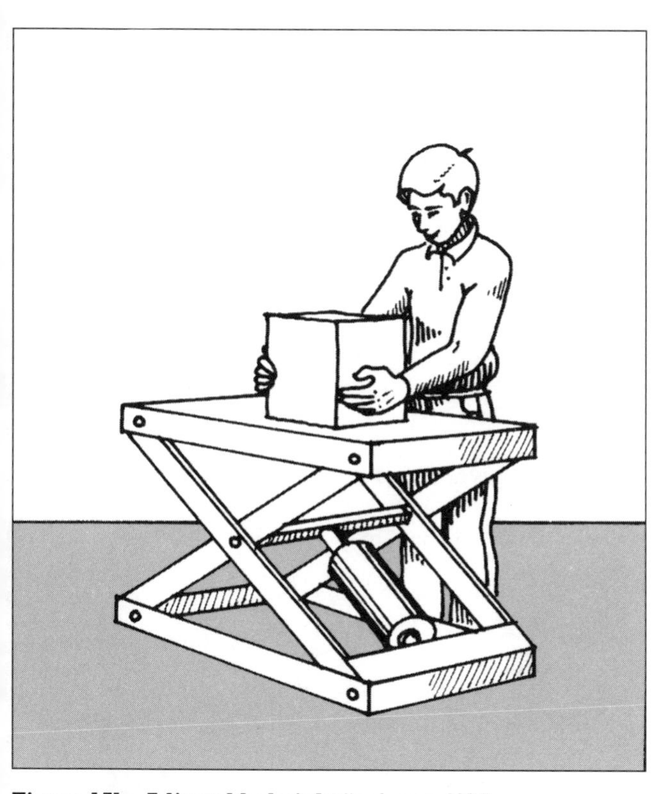

Figure 15b. Adjustable-height "scissors lift"

CHECKPOINT 16

Keep objects close to the body when carrying.

WHY

It is always a good solution to reduce the amount of manual carrying, but this is not feasible everywhere. If manual carrying is done carelessly, it increases both worker fatigue and the possibility of accidents.

Carrying an object close to the body minimizes the forward bending movement, thus reducing the risk of back injury, and neck and shoulder disorders.

By holding objects close to the body, carrying is easier and can provide a good forward view. This increases efficiency and reduces accidents.

HOW

1. Provide handles, grips or good holding points for the load carried. Approach the load as closely as possible and hold it firmly and near the body.

2. When lifting or lowering of a heavy load is involved, do this slowly in front of the body. Use the muscle power of the legs (not the back) and keep the back straight.

3. While carrying, keep the load near the waist. It is often useful to provide adequate aprons, as they minimize the risk of injuries from uneven or sharp parts of the loads.

4. Organize the carrying work so that it is done with minimal raising or lowering of the carried objects. For example, carry materials from a work surface to another work surface of the same height, or avoid putting materials on the floor by using stands or platforms of appropriate height.

5. When the load is heavy, consider the possibility of dividing it into smaller weights. If this is not possible, ask two or more people to carry the load, or consider the use of transport devices.

SOME MORE HINTS

- While the weight of the load should not be large, using pallets, trays, boxes or containers for carrying small items can reduce the number of trips. Wherever possible, the use of pushcarts and other mobile devices is naturally better.

- Consider the physical differences between different workers. Make sure that the weight and the frequency of loads are not excessive for the workers concerned.

- The worker may prefer to carry a load on the shoulder, the head or the back depending on its size and weight, and local custom. Try to find alternative means of easier carrying. If carrying is unavoidable, recommend the use of appropriate means of stabilizing the load, such as belts, an easy-to-carry container or a rucksack.

- Provide appropriate work clothes when the carrying task is frequent.

POINTS TO REMEMBER

When manual carrying of objects is unavoidable, lift and carry the object close to the body. This reduces fatigue and the risk of injury.

Figure 16a. Do manual lifting or lowering of a heavy load in front of the body, with the back kept straight and with stable foot positions, and use the power of the legs

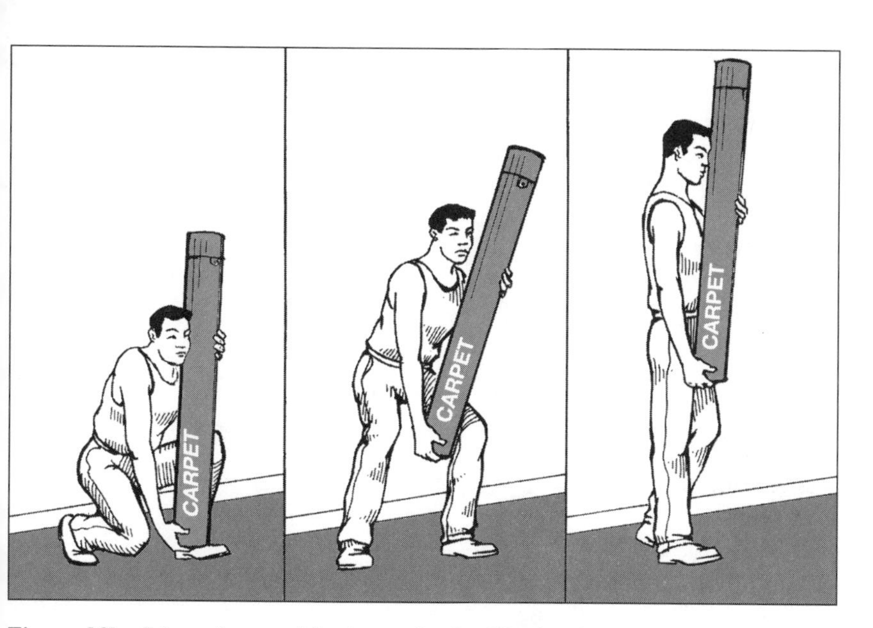

Figure 16b. A long heavy object can also be lifted using the power of the legs by keeping the object as close to the body as possible

Figure 16c. Handling grips adapted to the particular object to be carried can help ease the carrying work

CHECKPOINT 17

Raise and lower materials slowly in front of the body without twisting or deep bending.

WHY

While it is always better to use mechanical devices to raise or lower heavy materials, sometimes workers have to carry materials manually. Since lifting and lowering of materials are strenuous activities and can cause accidental injuries, they should be carried out by means of a correct lifting technique using the muscle power of the legs.

Raising and lowering materials slowly without twisting or deep bending of the body is more stable and reduces the risk of back injuries that may be caused by insecure movements.

Lifting materials in front of the body is less tiring for the worker, whose productivity is thus maintained.

HOW

1. Eliminate as much as possible the need to raise or lower materials manually. For example, move materials along work surfaces at the same height. If raising or lowering of materials is frequent, consider the use of mechanical means.

2. Wherever possible, minimize the height differences between the height of materials before or after carrying and the height of carrying. Use storage stands, side-tables, racks, platforms or workbenches on which materials can be placed near the carrying height.

3. Advise all workers about the correct techniques to raise and lower materials. The idea is to handle materials just in front of the body without twisting, to place the feet apart, to keep the back straight, and to raise or lower the load using the muscle power of the legs (not the back) and the grip of the hands.

4. Encourage the use of mobile racks, carts or trolleys to minimize the need for moving materials manually. The use of trays or pallets combined with these carrying devices can help reduce the frequency of raising or lowering materials manually.

SOME MORE HINTS

- Heavy loads should be raised and lowered slowly and steadily without jerking or snatching.

- Use a yoke, tongs, a sack-truck or similar device so that materials can be carried without lifting them much from the floor.

- A variety of manually powered lifting devices are available: hydraulic floor cranes, hydraulic lift-tables, lever hoists, chain hoists, etc. Try to use them instead of manual lifting.

- Avoid placing materials on the floor. Use platforms or stands that have some height.

POINTS TO REMEMBER

Raising and lowering materials slowly in front of the body reduces the worker's effort and the risk of injury.

Figure 17a. Lifting of heavy loads (i) from the floor and (ii) from a platform. Lifting from a platform is better than lifting from the floor

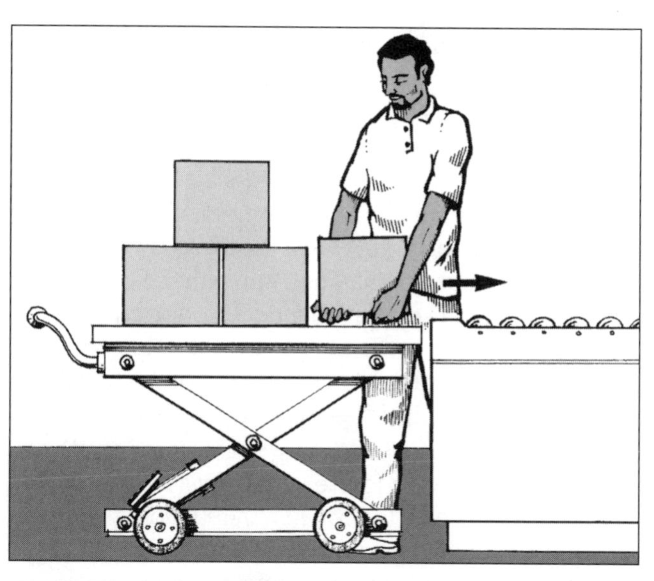

Figure 17b. Move materials along surfaces of the same height

CHECKPOINT 18

When carrying a load for more than a short distance, spread the load evenly across the shoulders to provide balance and reduce effort.

WHY

Carrying a load using both arms is more stable and safer than carrying the same load with only one arm. When the load can be divided into two parts, half the load for each arm can be carried far more easily than the double load carried by one arm.

Half the load carried by each arm gives good balance, allowing the worker to continue longer with less fatigue. A one-sided load can cause injuries and back, shoulder and neck disorders.

With less effort and less fatigue, carrying using both arms or both shoulders is safer than carrying the double load at one side.

HOW

1. Always consider whether the load can be carried using a pushcart, a trolley or some other wheeled device. This should be preferred to manual carrying, especially for heavy loads. However, if wheeled devices are not available or are impractical, then consider the following in manual carrying.

2. When a load of considerable weight has to be carried manually and cannot be divided, try to carry it with both hands in front of the body. Use a box, tray or appropriate container with handholds on both sides.

3. When carrying a load for some distance, consider if the load can be divided into two parts of almost equal weight. Then carry these two parts separately in separate hands. Use appropriate containers for carrying the divided loads. Containers of flexible shape such as bags with easy-to-grasp handles may be used.

4. If appropriate, try to use a yoke or a similar device by which the two separate loads can be carried together. Two special boxes or containers tied to both ends of the yoke can be carried while sometimes changing the shoulder to carry it. Tie the loads to the two ends of the yoke in such a way that the loads need minimum elevation when carried.

5. If the load is relatively heavy and an appropriate rucksack-style carrying device is available, you may carry the load on the back.

SOME MORE HINTS

- You can easily produce simple carrying devices like single- or double-handled tongs that require minimum elevation of the load.

- Share the load among two or more people when it is too heavy or too large to be handled by a single person. Even in this case, the use of a cart or some other wheeled device is preferable.

POINTS TO REMEMBER

Distribute the manually carried load between both arms rather than carrying it with one arm only. If the load is heavy, however, use a cart or some other wheeled device.

Figure 18. A yoke or a similar device is useful for carrying two separate loads for some distance while keeping balance and minimizing the lifting or lowering work

CHECKPOINT 19

Combine heavy lifting with physically lighter tasks to avoid injury and fatigue and to increase efficiency.

WHY

Manual lifting of heavy loads is tiring and a major source of back injuries. If this cannot be replaced by the use of a wheeled device or mechanical transport, it is better to combine the heavy lifting with lighter tasks. The idea is to avoid concentrating unfavourably heavy tasks on a few workers.

Combining heavy lifting with lighter tasks reduces fatigue, as well as the risk of back injuries. This helps increase the worker's overall productivity.

If workers are trained to perform multiple tasks, it is much easier to find a substitute worker in the case of absence of one worker due to illness or leave.

HOW

1. Reorganize job assignments so that workers who perform heavy lifting tasks perform lighter tasks as well.

2. Introduce job rotation and group work in order to prevent biased concentration of strenuous tasks on selected workers.

3. If manual lifting of a heavy load is unavoidable, try to share the load by having two or more people carry it together.

4. For similarly strenuous tasks, consider job assignments so that these strenuous tasks can be shared by a group of people on a rotating basis.

SOME MORE HINTS

- Physically demanding tasks such as heavy lifting always require frequent breaks. Allow sufficient breaks for recovery from fatigue and for better productivity. Requiring frequent breaks as part of the work schedule may give you an impetus to combine the heavy tasks with lighter tasks.

- Alternating tasks is often far less tiring, and thus improves the worker's motivation and productivity.

POINTS TO REMEMBER

Avoid repeated heavy lifting all the time. Combine heavy lifting with lighter tasks in order to reduce fatigue and increase efficiency.

Figure 19. (i) and (ii) Combine physically heavy work with lighter tasks. This reduces fatigue and increases efficiency

CHECKPOINT 20

Provide conveniently placed waste containers.

WHY

Waste, scrap and spills of liquid on the floor not only represent a loss of material and a hindrance to the smooth flow of production, but are also an important cause of accidents.

Good housekeeping is difficult without providing waste containers in convenient places.

Conveniently placed, easy-to-empty waste containers help to create free space and reduce cleaning costs.

HOW

1. Purchase or construct waste containers suitable for each type of waste: open box-style or cylinder containers or bins for scrap and rubbish (large enough for the type of waste); closed containers for liquids; and appropriate racks or platforms for longer or larger waste (such as wood splints, metal rods, etc.).

2. Put wheels under the waste containers so that the waste can be carried to the place of disposal frequently and easily.

3. If oil or other liquid is spilt from machinery or from transport systems, construct removable trays underneath the object.

4. Consult workers about the best way to empty waste containers at appropriate intervals. Assign the responsibility of emptying waste to one person or rotate the emptying task among a group of workers. The idea is to include waste disposal as one standard activity of the work process.

SOME MORE HINTS

- Simple metal or plastic containers placed at each work area can often help maintain good housekeeping.

- A vacuum cleaner is a good, temporary waste container for small, dry particles. Wet waste requires a special vacuum cleaner (consult your dealer).

- Heavy waste can be handled more efficiently if the container can be opened at the appropriate height, e.g. from the side of the container.

- Waste materials stored in containers remain relatively clean, suffer less deterioration and can be easily located when needed.

POINTS TO REMEMBER

Well-organized waste storage is necessary for good housekeeping. Waste thus stored can be recycled.

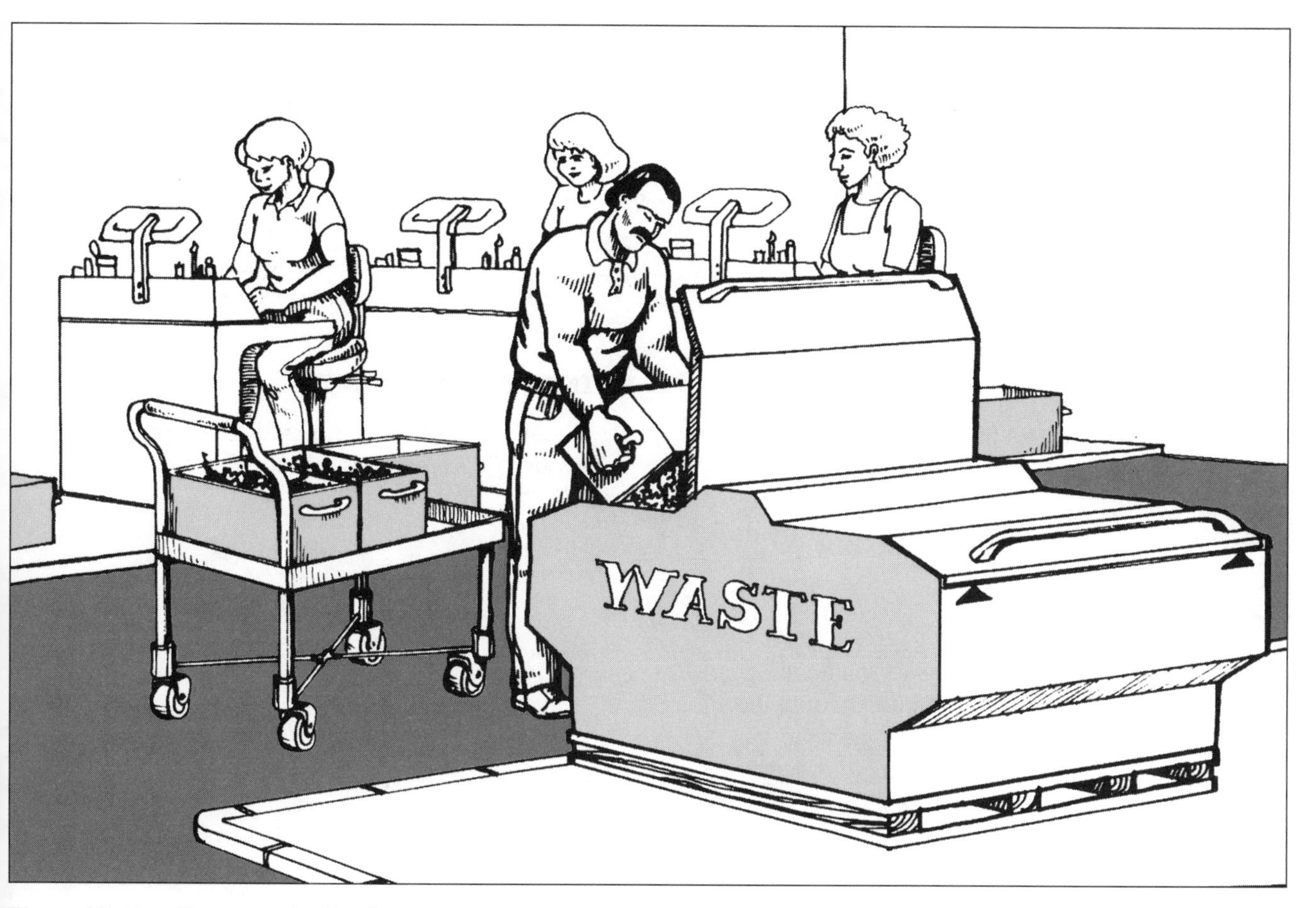

Figure 20. Provide conveniently placed, easy-to-empty waste containers

CHECKPOINT 21

Mark escape routes and keep them clear of obstacles.

WHY

It is important to keep escape routes always clear of obstacles.

Escape routes, if rarely used, tend to be neglected and thus obstructed by piled-up materials, wastes or equipment. It is too late to start clearing the escape routes after a fire breaks out.

In an emergency, people get upset and may even panic. Therefore, escape routes must be easily recognizable and simple to follow.

HOW

1. Make sure that at least two exit routes are present in each work area. Take into account the possibility of a fire breaking out near an exit route. Check legal requirements for escape routes.

2. Mark each escape route on the floor unless it is explicitly clear that it is the escape route (as in the case of fenced aisles or corridors). Clearly indicate the emergency exit. Where the emergency exits are not in sight, clearly post the direction of the nearest emergency exit.

3. Firmly establish the practice of placing nothing on escape routes and of keeping them clear of obstacles at all times.

4. Use fences, hand-rails or screens to create a clear space around the emergency exits and to allow easy access to them. Put fences or hand-rails along the escape routes when these tend to be obstructed by stacked goods.

SOME MORE HINTS

- Check if the escape routes are easy to recognize and follow at all times when workers are present, e.g. during an evening or night shift, or in the case of a sudden power cut.

- Organize evacuation drills at appropriate intervals, and use the drills as an opportunity to ensure that escape routes are free from obstacles. Too frequent evacuation drills, however, can be a problem. Rather, have a designated person (or team) check the escape routes regularly.

- Provide storage shelves, pallets, racks or waste bins near the work areas or passageways where materials and work items tend to pile up. This helps to keep the escape routes clear at all times.

POINTS TO REMEMBER

Clear and easily recognizable escape routes can save your life. An emergency can happen at any time.

Figure 21. Mark escape routes and keep them clear of obstacles

Hand tools

CHECKPOINT 22

Use special-purpose tools for repeated tasks.

WHY

Special tools adapted to the particular operation greatly improve productivity. These tools make the operation easier and therefore much safer.

Special tools can usually be purchased or made at low cost. Since productivity increases as a result, the benefits are far greater than the cost incurred.

HOW

1. Use special-purpose tools to do exactly the right job with the best quality and the least effort. Use precisely the right type, size, weight and strength of screwdrivers, knives, hammers, saws, pliers and other hand tools.

2. If the task requires frequent strenuous effort, use motor-powered tools. Various types are available. These motor-powered tools are not only more efficient but can perform tasks which are impossible for manual operation. Workers' fatigue is far less.

3. Arrange for "homes" for tools not in use, and for their regular maintenance.

4. Instruct workers about the correct use of the tools. Have them ask for repair or replacement when the tools are damaged or worn out.

SOME MORE HINTS

- Tool cost has three components: purchase (or production), maintenance, and use of utilities. Most non-powered hand tools cost less than the hourly labour cost. Even powered hand tools usually cost less than 20-50 times the hourly labour cost. Yet tools are used for 1,000-2,000 hours per year over a number of years. For example, a powered screwdriver costing 50 times the hourly labour cost, used for 1,000 hours a year for 5 years, actually costs only 1/100th of the hourly cost (50/5,000=1/100).

- Non-powered tools tend to need 0-5 hours of maintenance per year; powered tools 10-100 hours a year. Thus even at 50 hours' maintenance, in a year they cost a small fraction (say 1/30th) of the hourly labour cost.

- Even a relatively expensive tool (e.g. a powered screwdriver), including maintenance and utility expenses, costs only about 3 per cent of the labour cost per hour. Compare this with the increase in productivity by the use of the tool, which is surely far more than 3 per cent. Also consider better-quality output and less stress for workers.

POINTS TO REMEMBER

Special-purpose tools are an extremely worthwhile investment. They are cheap, and improve productivity and safety to a great extent.

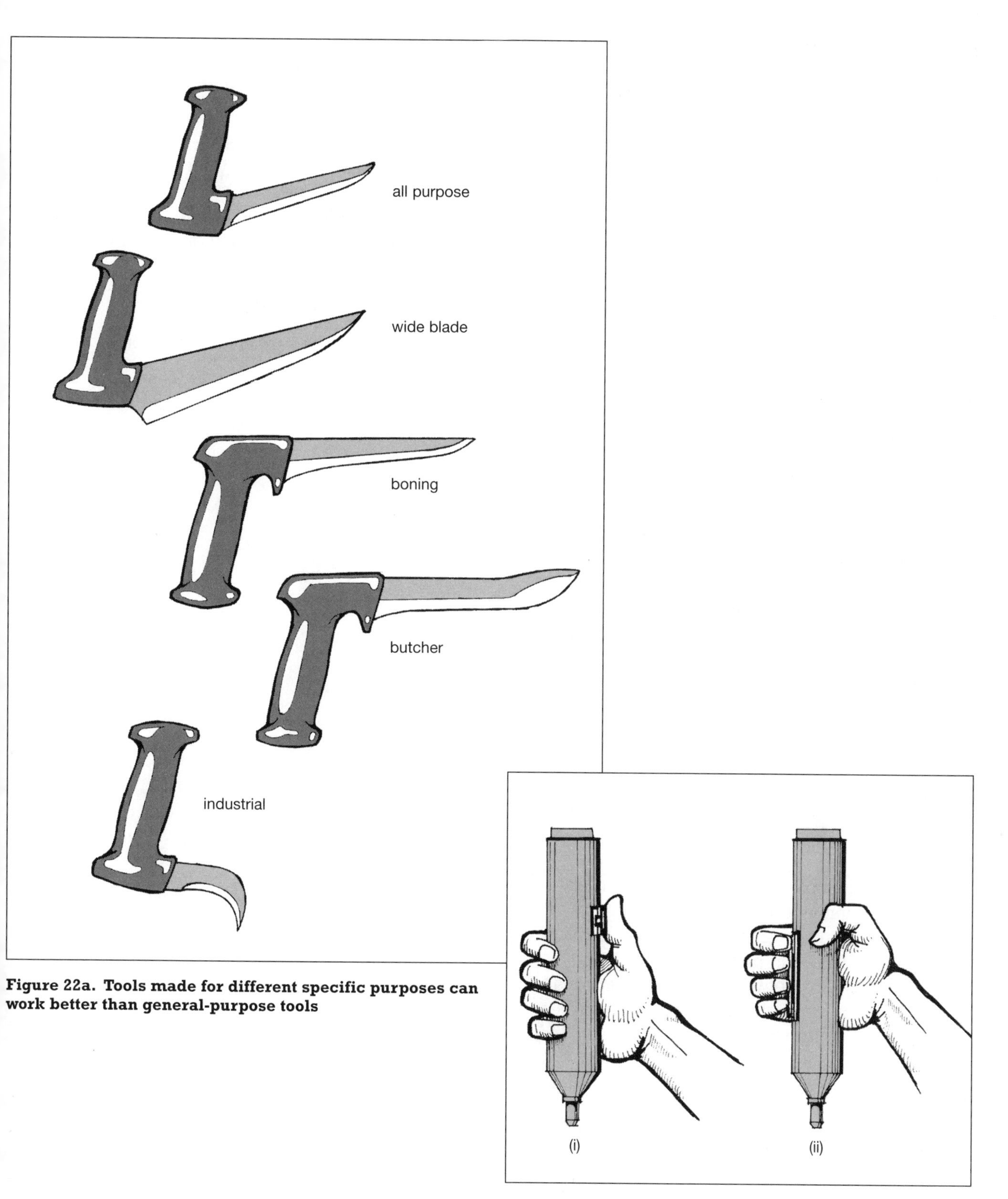

Figure 22a. Tools made for different specific purposes can work better than general-purpose tools

Figure 22b. Thumb-operated and finger-strip-operated pneumatic tools. (i) Thumb operation results in over-extension of the thumb. (ii) Finger-strip control allows all the fingers to share the load and the thumb to grip and guide the tool

CHECKPOINT 23

Provide safe power tools and make sure that safety guards are used.

WHY

Power tools are efficient but usually more dangerous than non-power tools; the greater the energy, the greater the danger. Nevertheless, safe power tools are available. There is no need to use unsafe tools.

Safe tools are more productive.

HOW

1. Order power tools only after safety specifications have been examined. Three important points are: protection against power transmission and points of operation; prevention of unintended activation of controls; and easy operation with secured grips.

2. Compare the guards provided for the power tools you are going to purchase with those of other kinds or of similar tools. Not only dealers but also your fellow workers can help.

3. Check if the guards are sufficient to protect workers, and are used.

4. Guards should not interfere with work, or people will remove them.

5. Check if multiple means are used in order to prevent unintended activation of controls. For example: controls are not protruding towards the outside of the equipment; there should be sufficient space between controls; controls are recessed or covered with a barrier; controls require more than minimal force to activate them; an interlock, a key or an enabling control is used for power switches.

SOME MORE HINTS

- Organizations, not workers, should buy and maintain work tools. Workers usually do not have the technical knowledge to know which is the best tool and may also not be able to pay for such a tool.

- Consider the whole line of action from fetching the tool, setting it in motion, going from one operation to another, and putting it away. Is the worker safe throughout?

- There are two classes of guard: (1) equipment guards, and (2) people guards (also called personal protective equipment). Do not forget people guards. Make such guards (gloves, aprons, shields, etc.) available.

POINTS TO REMEMBER

A safe person is a productive person. A safe tool is a productive tool.

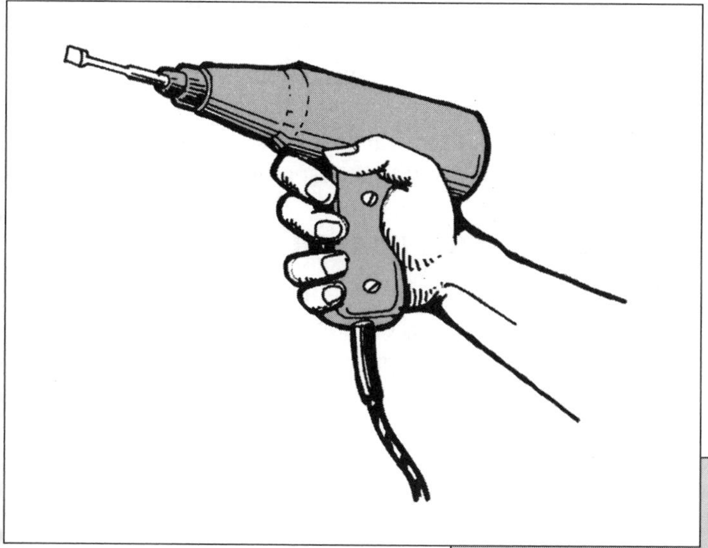

Figure 23a. An example of a tool with a trigger that is long enough, allowing a firm grip by the hand

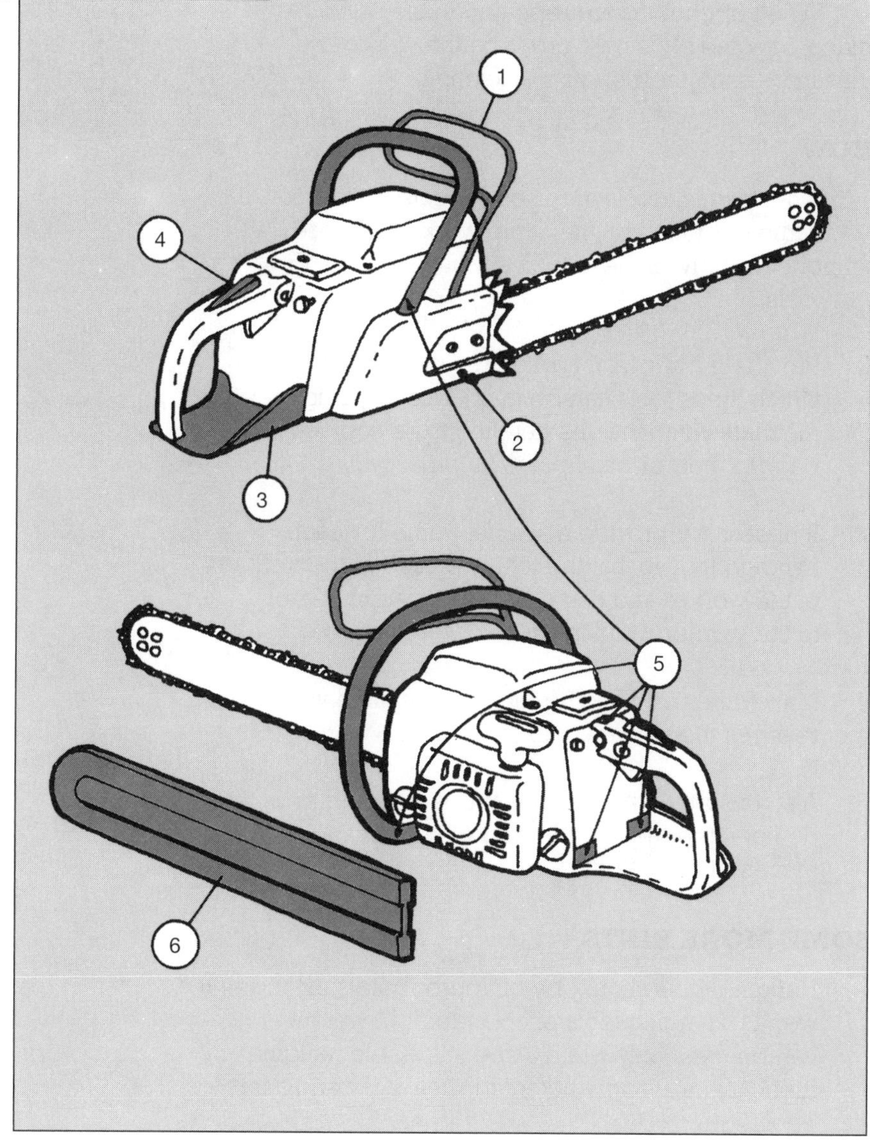

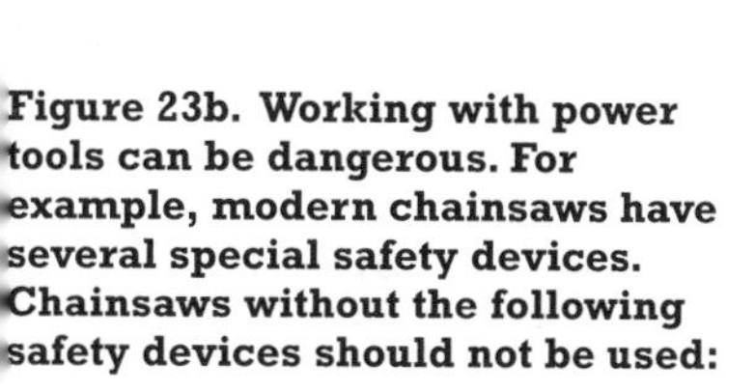

Figure 23b. Working with power tools can be dangerous. For example, modern chainsaws have several special safety devices. Chainsaws without the following safety devices should not be used:

1. **Front handle guard with chain brake (protects left hand and stops saw chain during kickback)**
2. **Chain catcher (catches the saw chain if it breaks)**
3. **Rear handle guard (protects right hand)**
4. **Throttle control lock-out (prevents saw chain from starting to move unexpectedly)**
5. **Anti-vibration devices (prevent vibration diseases of the hands)**
6. **Guide bar cover (avoids injuries during transport of chainsaw)**

CHECKPOINT 24

Use hanging tools for operations repeated in the same place.

WHY

Hanging tools can be grasped easily near the point of operation. You can save the time needed to put the tool down and to pick it up again. The time grasping the tool is shorter and the worker's fatigue is less.

Hanging tools are always easy to find. You need not create a "home" for them, such as a tool-rack or a side table. You can save space in this way.

When operations are repeated in the same place, hanging tools help organize the workplace, thus increasing the worker's efficiency.

HOW

1. Check which tools are used repeatedly for the same operation by the same worker. Choose one or more of these tools for use as hanging tools.

2. Provide a horizontal frame over the worker from which these tools can be hung. Use a spring mechanism so that the hanging tools can go back to their original place automatically.

3. If necessary, provide a special frame for each hanging tool so that the tool is placed at the front of the worker and can easily be brought nearer to the worker while in use.

4. Make sure that hanging tools are within easy reach of the worker.

5. Also ensure that, when not in use, hanging tools do not interfere with the worker's arms and movements.

SOME MORE HINTS

- Hanging tools should be of appropriate size and weight. Hanging tools of considerable weight can be used only when a special, stable hanging mechanism is constructed for their easy and safe operation.

- If hanging tools are to be used by different workers, make them adjustable to the workers' varying levels of hand reach.

- In the case of wear or breakage, hanging tools should be easy to replace and maintain.

POINTS TO REMEMBER

Hanging tools provide a good solution for storage and easy operation, while increasing the worker's efficiency and productivity.

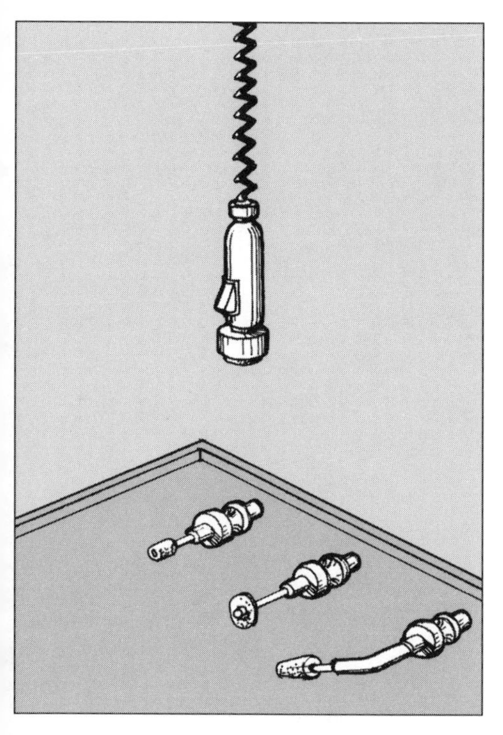

Figure 24a. A handy and safe power tool that has operating parts exchangeable for each specific purpose can save time and effort

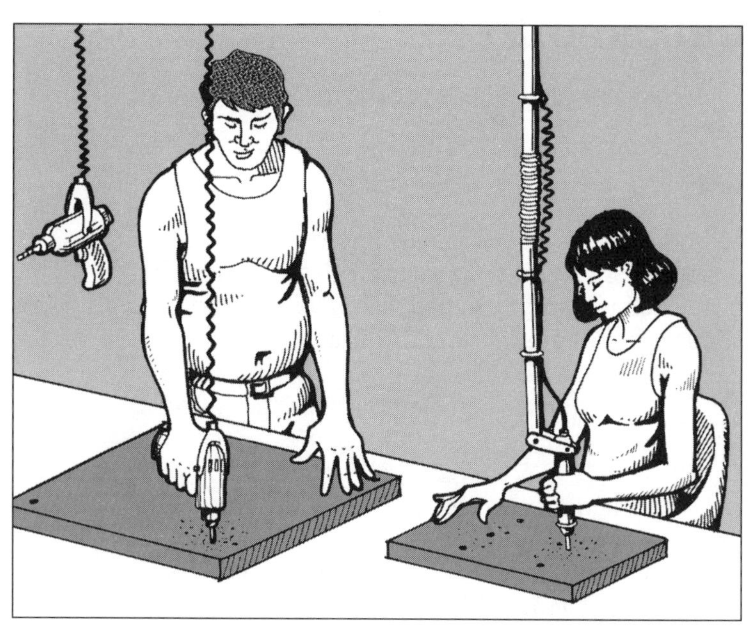

Figure 24b. Hanging tools within easy reach of workers are suited to repeated operations

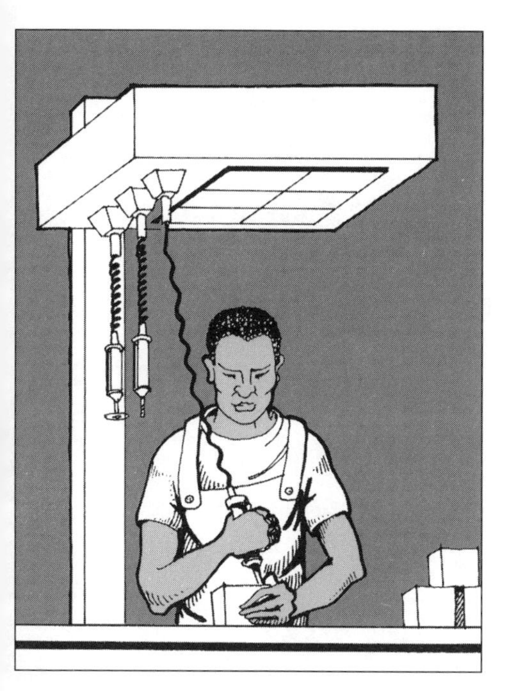

Figure 24c. Provide a special overhead frame from which tools can be hung by means of spring mechanisms

Figure 24d. Placement of tools in accordance with frequency of use

CHECKPOINT 25

Use vices and clamps to hold materials or work items.

WHY

Manual operations greatly improve when the materials or work items are firmly fixed. Vices and clamps allow workers to use different sizes and shapes of workpieces steadily during the work.

The use of vices and clamps allows the workers to use both hands.

Vices and clamps also reduce accidents, as they prevent slippage of material, reduce the need for maintaining a bad posture and provide better control over the work item.

HOW

1. Select the appropriate vices or clamps, considering the sizes and shapes of work-pieces.

2. If possible, make the location of the vice or clamp fixture adjustable on the work surface.

3. If the task requires the worker to access the work-piece from different directions, select clamps that can rotate.

4. Locate vices and clamps to allow workers to perform work in a natural posture or position. The working height should be a little below elbow level.

SOME MORE HINTS

- Select a vice or a clamp that minimizes the force required to secure the workpiece in it.

- Make sure that the vice or clamp has no sharp edges.

- Select a vice or clamp that allows the worker to know when the workpiece is secured in place, without damaging the workpiece.

POINTS TO REMEMBER

A work item secured in a vice or clamp is a safe work item.

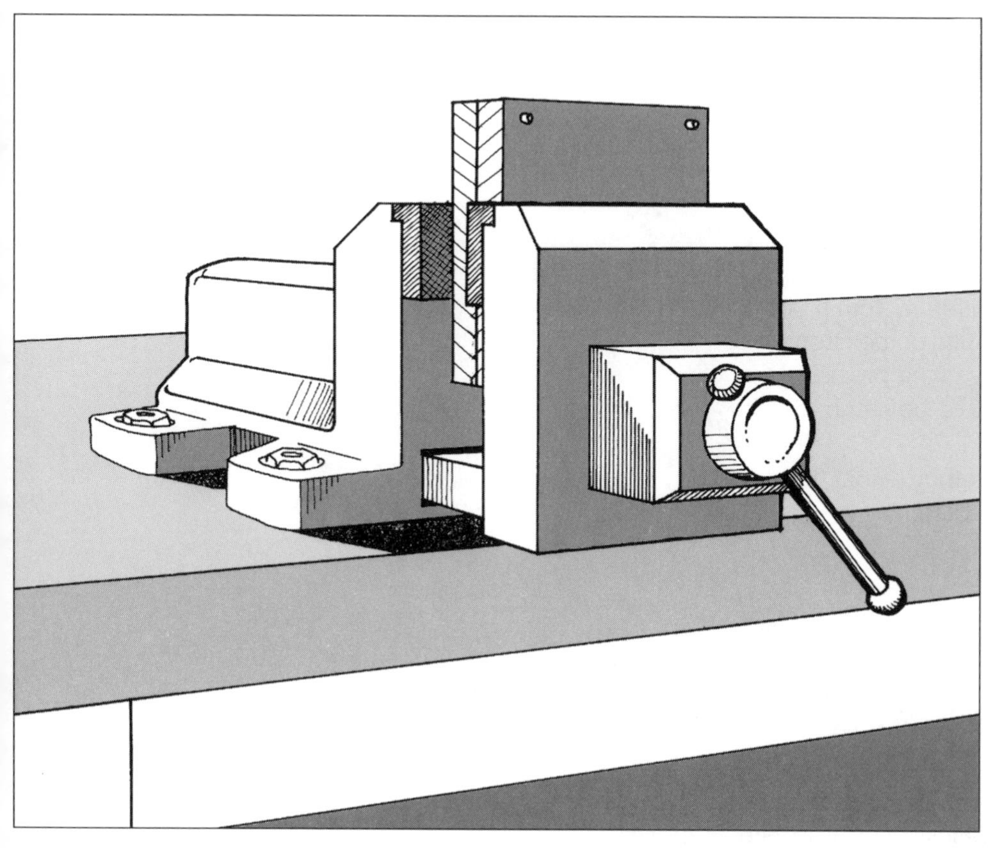

Figure 25a. Use a vice or a clamp that can steadily hold the work item at the appropriate height

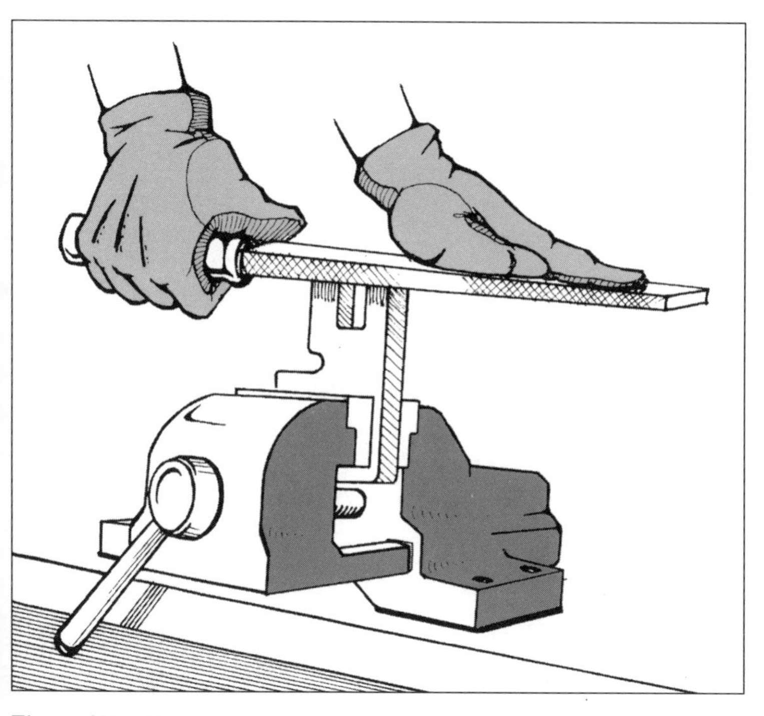

Figure 25b. The use of vices or clamps allows the worker to use both hands for productive work

CHECKPOINT 26

Provide hand support when using precision tools.

WHY

Accuracy of tool operation or precision work depends much on the stability of the hand doing the work. The precision grip is different from a power grip and requires about one-fifth of the strength of the power grip. The accuracy of the precision work is affected by slight movements of the hand.

Hand support reduces tremor (slight trembling). Reduced tremor increases accuracy.

HOW

1. Provide a support near the point of operation so that the hand (0.6 per cent of body weight) or the hand and the forearm (2.8 per cent of body weight) can be supported during work.

2. Try out various positions and shapes of the hand support to get the best results. If appropriate, provide an adjustable support.

3. If appropriate, place the precision tool on a support as much as possible. Artists have been using steady rests for centuries.

SOME MORE HINTS

- Minimize forceful exertions with the hand as the arm muscles which control the hand are very sensitive to tremor. For example, surgeons should not carry a suitcase for 24 hours before performing an operation.

- Shield the front of some tools (e.g. a soldering iron). The shield reduces the impact of flying objects from the work, and acts as a support for the hand. The shield will prevent the forward slip of the hand, and the operator can grip the tool closer to the work, giving better control of the tool tip.

POINTS TO REMEMBER

To increase accuracy, support the precision tool or the hand doing the work, or both.

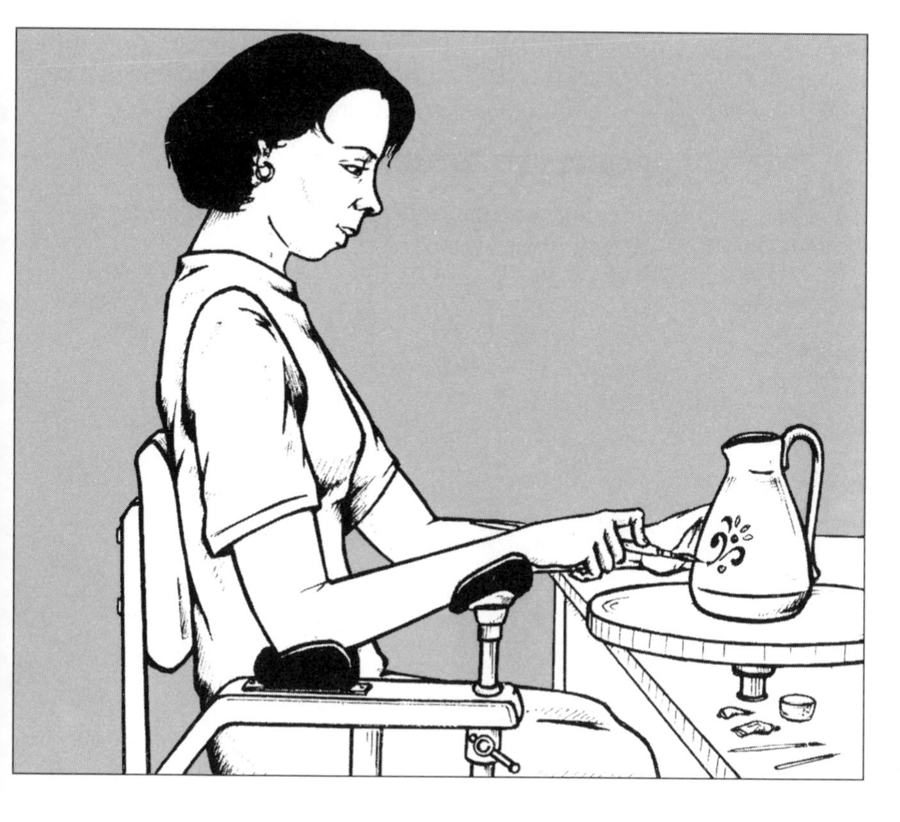

Figure 26a. A hand or forearm support near the point of operation increases the efficiency of precision work

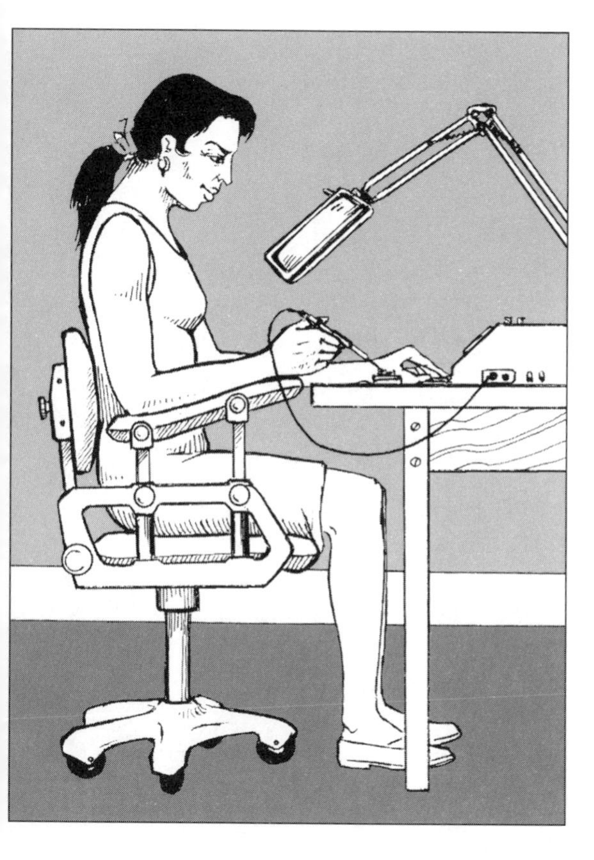

Figure 26b. Try out various positions and shapes of the hand support to get the best results

CHECKPOINT 27

Minimize the weight of tools (except for striking tools).

WHY

Tool weight often fatigues the user, thus reducing productivity.

Except for striking tools (hammers, axes), tools of the minimum weight are easier to handle and allow precise operations.

Lighter tools are easier to store and simpler to maintain.

HOW

1. Choose tools suited to the purpose but of minimum weight.

2. There are various ways to minimize the actual weight that has to be held by the hand. For example, support the tool on a steady rest; this also improves accuracy.

3. If appropriate, slide the tool along a surface (the surface supports the tool).

4. Suspend the tool on a balancer above the tool's centre of gravity. Typically the balancer pulls upward with slightly more than the weight of the tool (e.g. 2.1 kg on a 2 kg tool). When the tool is released, it goes up and out of the way (but still within reach).

SOME MORE HINTS

- Work with the tool close to rather than away from the body. In this way, the actual force required to handle the tool is less. For example, a 2 kg tool held by the hand at the end of an arm that is 70 cm long exerts a 140 kg/cm rotation force (torque) about the shoulder, while the same tool held at only 35 cm away from the shoulder exerts a 70 kg/cm rotation force. The worker feels as if he or she is holding a much lighter tool.

- Use tools with the tool handle below the tool balance point (centre of gravity). If there is little effective tool weight and if the tool balance point is difficult to find, then you can locate the appropriate tool handle position (that minimizes tool action force) by making some trials.

POINTS TO REMEMBER

Lightweight tools reduce fatigue, allow better accuracy and increase productivity.

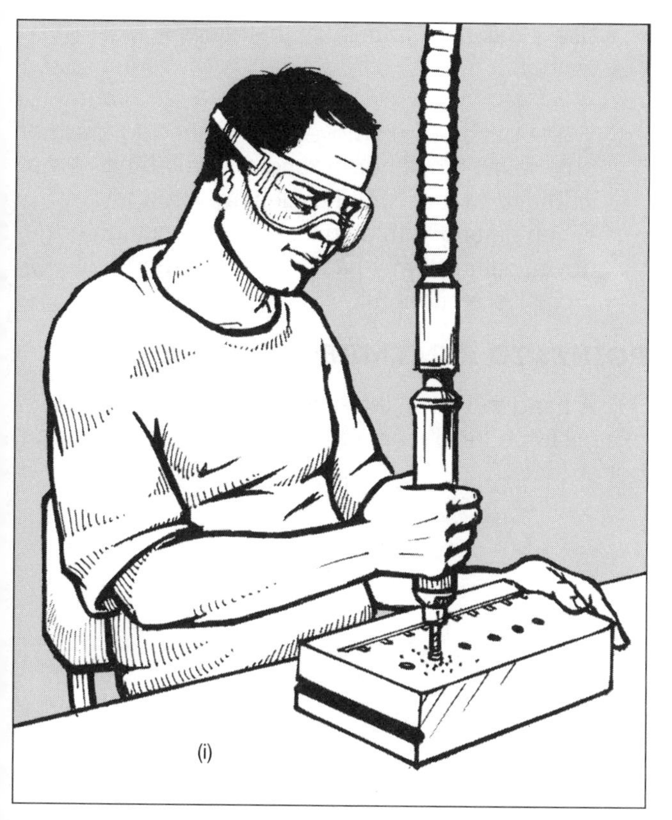

Figure 27a. (i) and (ii) Suspension of the tool above its centre of gravity can make the tool work more easily and effectively

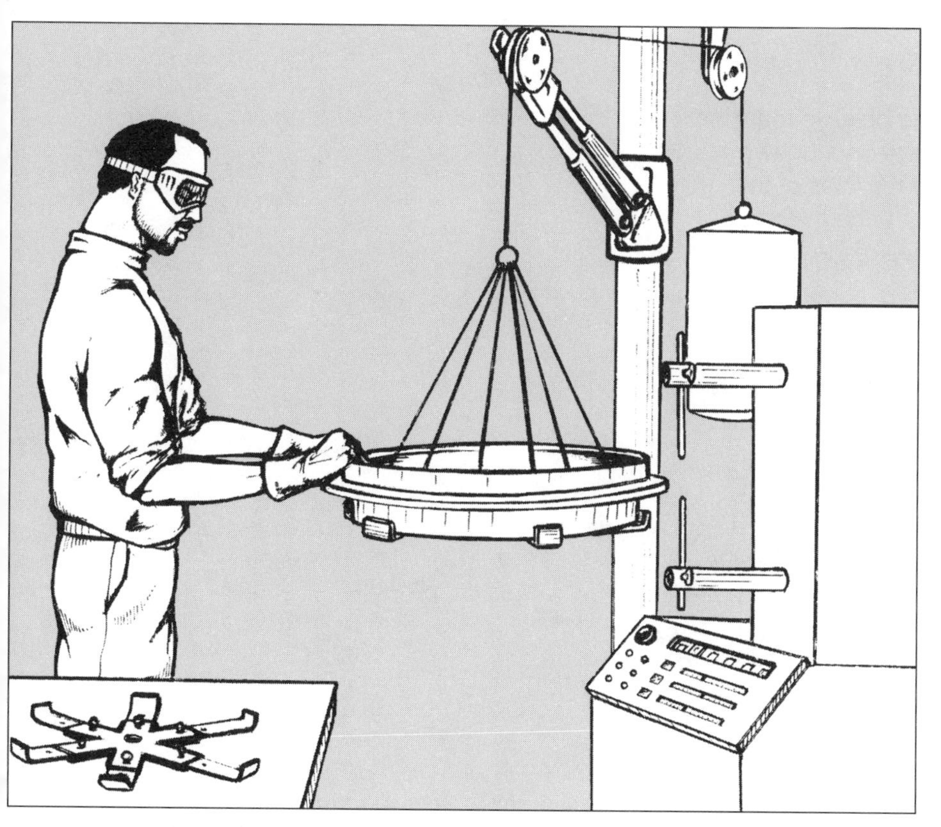

Figure 27b. Balancing mechanisms can also be used to minimize the weight of tools and work items together

CHECKPOINT 28

Choose tools that can be operated with minimum force.

WHY

In operating hand tools, small muscles in the fingers and hand are frequently used. If excess force is needed, these muscles get tired very easily.

Since delicate movements are essential in tool operation, even slight muscle fatigue reduces output.

Repetitive tool operations requiring excess force cause neck, arm and wrist disorders which can be very painful.

HOW

1. Avoid tools that require excess finger force. The muscles for moving fingers are particularly affected by excess force. For example, use trigger strips rather than trigger buttons, as fingers combined are stronger than individual fingers. Consult experienced workers when choosing good tools.

2. Choose tools that allow the use of large muscles. For example, a protective flange on a power screwdriver permits the large muscles of the forearm to resist the tool thrust force instead of squeezing by the small muscles of the fingers.

3. Minimize the time of muscle use. For example, when drilling a hole in sheet metal with a portable drill, use the muscles to start the drill but once started the drill trigger need not be held. The drill should shut off either after a predetermined time or when the drill has penetrated. (Note, however, that a trigger lock can be very dangerous when the tool is dropped, as it does not stop running.)

4. Whenever appropriate, use motor-powered tools. Motors extend human capability and do not get tired.

SOME MORE HINTS

- Use springs (not muscles) to open scissors, pliers, clippers, etc.

- Use a balancer to reduce the effect of tool weight.

- When push or pull movement is needed, push or pull below the shoulder and above the hip; within this range the muscles are strongest. When cutting with a knife, keep the cutting edge down; cutting away has twice the strength of cross-body motions.

POINTS TO REMEMBER

A hand tool is an extension of the hand. Avoid excess force for delicate tool operations by choosing better tools.

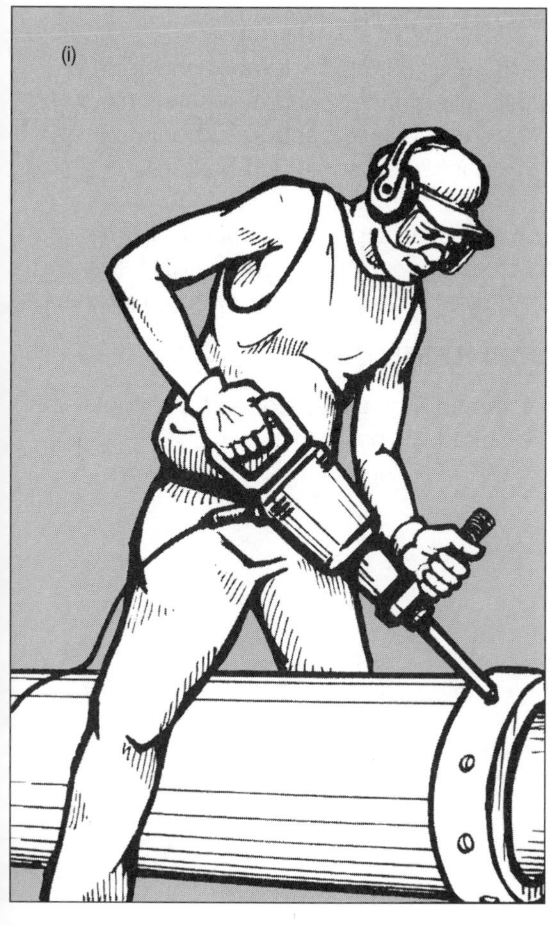

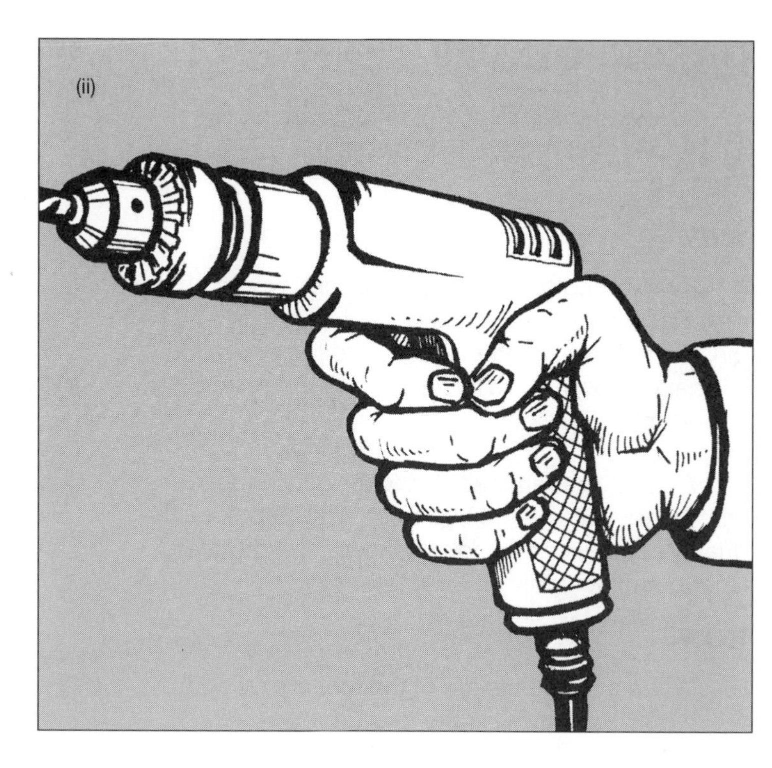

Figure 28. (i) and (ii) Steady grips of power tools help reduce the required force of operation

CHECKPOINT 29

For hand tools, provide the tool with a grip of the proper thickness, length and shape for easy handling.

WHY

Every hand tool has two ends; one works on the material and the other on the hand. The grip end must be adapted to the hand and to the operation. Its shape as well as its thickness and length are important.

A good grip allows the worker to use the tool with firmer control and less force. This improves the quality of the work being produced, and reduces fatigue and accidents.

HOW

1. When a single handle of the tool is grasped by the whole hand (i.e. the four fingers reach around the handle and are "locked" by the thumb over the first finger), ensure that the handle diameter is 30-40 mm. For double-handled tools, the initial span (that exists before using the tool) should be less than 100 mm and the closed span 40-50 mm with a handle diameter thick enough to cause no pain.

2. In the case of a hook grip (briefcase-style, with the four fingers acting as a group but the thumb passive and relaxed), or in the case of an oblique grip (golf-club-style, with the thumb pointing along the tool axis to improve precision), use a handle diameter of 30-55 mm.

3. Make sure that the handle length is at least 100 mm; 125 mm is more comfortable. Use a handle of at least 125 mm if the hand is "enclosed" (e.g. a saw) or if a glove is worn.

4. Check whether the tool size provided is suitable for the individual worker. Tools are often designed for male hands; for tools to be used by women, you may need to buy from vendors who furnish smaller sizes.

5. Check whether the wrist is able to stay in the neutral (handshake) position while the tool is used. For example, a pistol-style grip may be a good solution.

SOME MORE HINTS

- Make the grip usable by either hand. This is because using either hand alternately can help to reduce cumulative trauma, and because about 10 per cent of people are left-handed.

- Note that gloves increase the hand size. It is therefore necessary to try out the grip size and its hand clearance by using it with gloved hands.

POINTS TO REMEMBER

A tool should fit you like clothing. Use tools with a grip size suited to you.

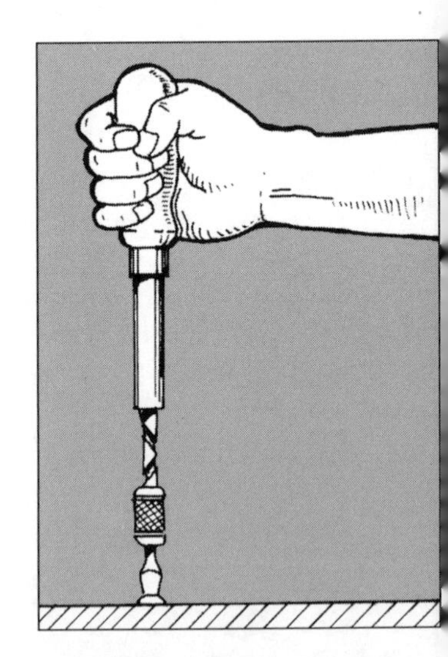

Figure 29a. Alternative tools can be used to reduce mechanical stress. For example, a screwdriver equipped with a ratchet reduces mechanical stress on the hand, as well as the amount of force necessary to drive and remove screws

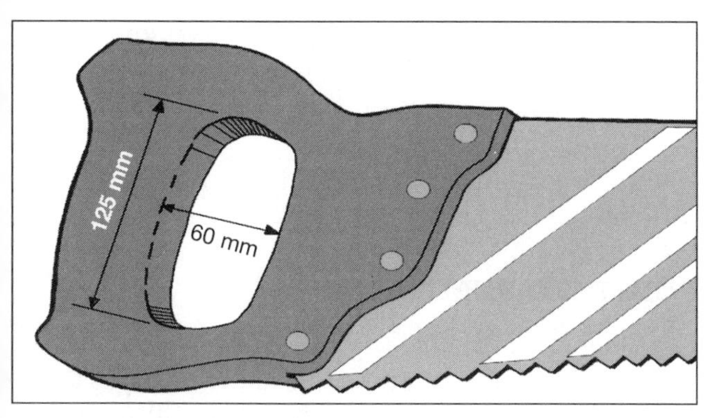

Figure 29b. The tool grip should be of the proper thickness, length and shape

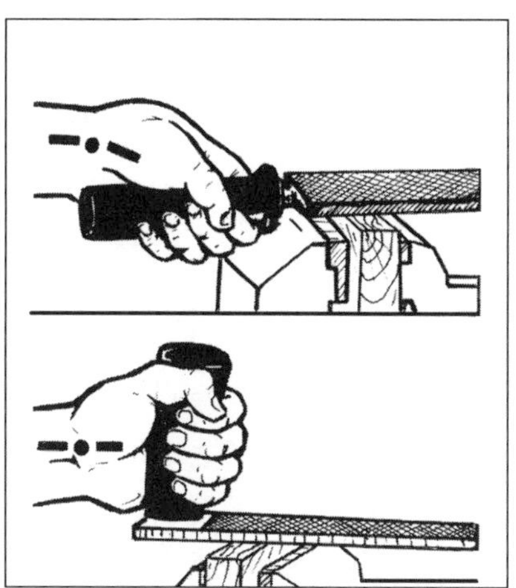

Figure 29c. For firm and safe tool operations, allow the four fingers to reach around the grip, enabling the thumb to come over the first of them

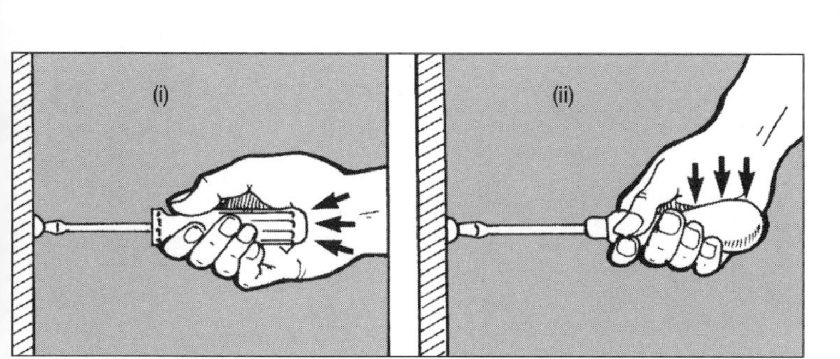

Figure 29d. The handle in (ii) reduces mechanical stress by distributing force over a larger area of the hand than in (i)

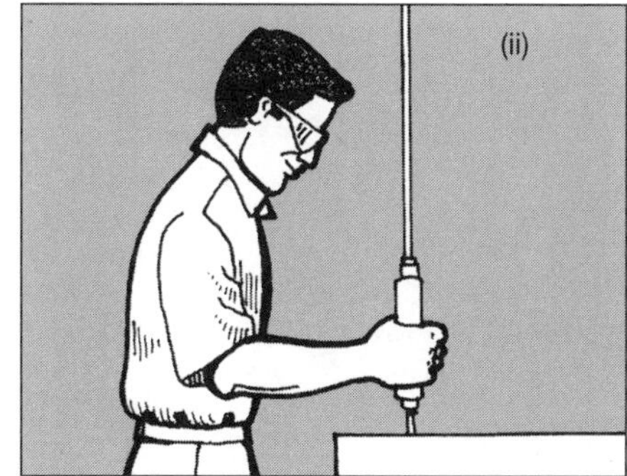

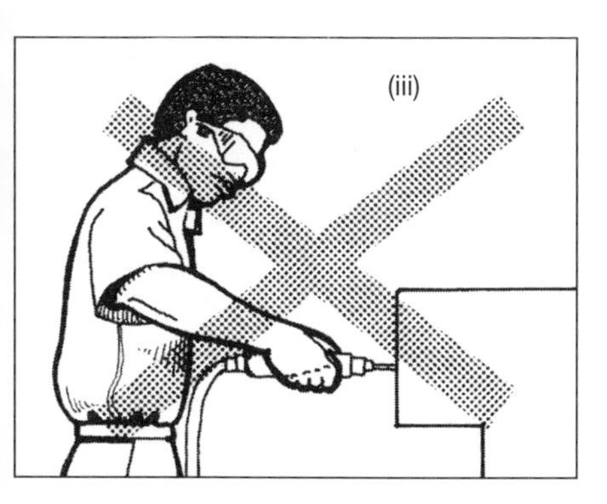

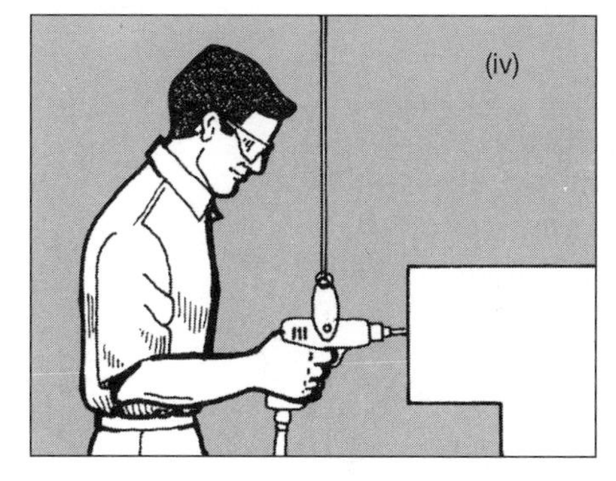

Figure 29e. A tool should be used (i) and (ii) on a horizontal surface at elbow height, or (iii) and (iv) on a vertical surface below knuckle height. Suspension of the tool can help you get a better grip

CHECKPOINT 30

Provide hand tools with grips that have adequate friction or with guards or stoppers to avoid slips and pinches.

WHY

Slips of the hand or pinches while using tools cause injury. Slips and pinches can be prevented by improving the tools.

Loss of control of the tool may damage parts. Fear of slips and pinches reduces the quality of work.

HOW

1. Reduce tool rotation in the hand by using grips that have a non-circular cross-section, and by using grip surface material with a good coefficient of friction (e.g. vinyl, rubber, soft plastic).

2. Use a wedge-shaped tool (with a change in cross-section) to reduce the forward movement of the hand and to allow more force to be exerted.

3. Use guards or stoppers at the front (e.g. for knives and soldering irons) to act as a shield against slips, as well as to reduce hand movement and allow more force.

4. Use pommels (shields at the rear of the tool grip) to prevent loss of the tool and to make movement of the tool towards the body easier.

5. Choose tools with grip shapes that do not produce pinches.

SOME MORE HINTS

- Tool surfaces tend to become slippery after a period of work because of sweat, oil, etc., on the hand. Grips should be covered with good friction material. Guards against slipping are particularly important when strong force is exerted when using the tool.

- If there is a guard to prevent slipping, you can hold the tool further forward and improve accuracy.

- Open two-handled tools (such as scissors or pliers) with a spring (that is, the tool is "normally open") are often useful.

- Occasionally the tool should rotate in the hand; in this case a circular cross-section is useful.

POINTS TO REMEMBER

Guards on grips to prevent slips and pinches can reduce accidents, as well as improve work quality. Purchase or choose hand tools with such grips.

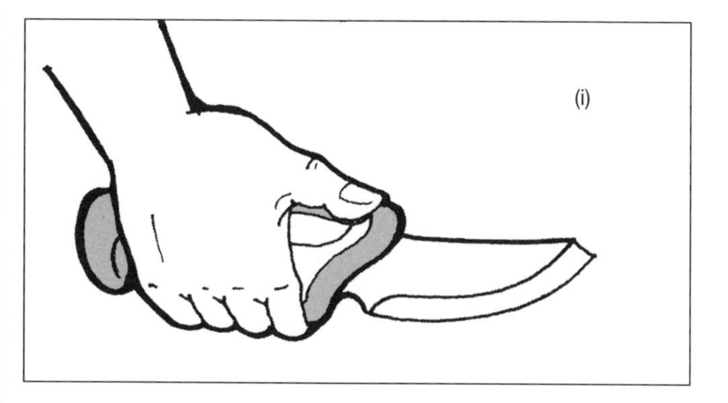

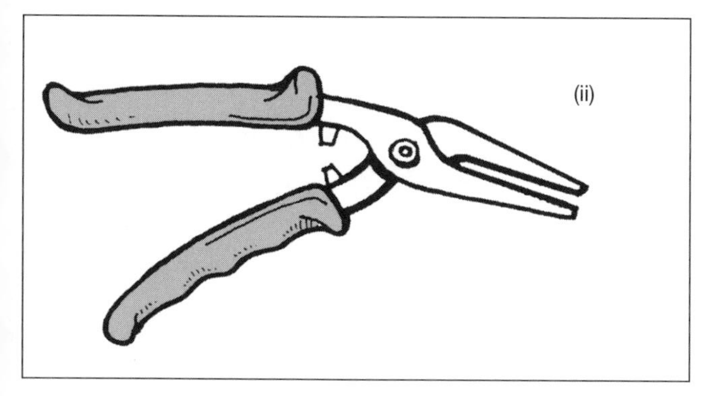

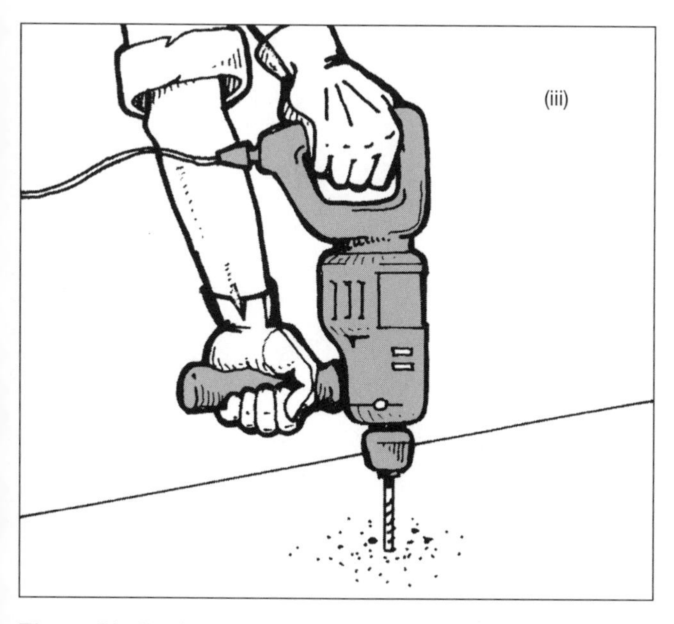

Figure 30. (i), (ii) and (iii) Grips that prevent the forward movement of the hand allow safe and effective tool operations

CHECKPOINT 31

Provide tools with proper insulation to avoid burns and electric shocks.

WHY

When using hand tools, workers tend to concentrate hard on the point of operation and forget the danger of burns and shocks. Burns and shocks while using tools are preventable.

Materials with low thermal conductivity have low electrical conductivity. Thus protection against burns (and freezing) also protects the worker from electric shocks.

HOW

1. Use grip surface material with low thermal conductivity, such as rubber, wood or plastic. Metal has high thermal and electrical conductivity and can be dangerous.

2. For metal handles, even a thin layer of plastic (e.g. a sleeve) can greatly reduce thermal conductivity and increase grip comfort.

3. In the case of electrically powered hand tools, use grounded plugs and double-insulated tools (where the handle is insulated from the power).

SOME MORE HINTS

- When the danger of burns or electric shocks while using the tool is real, use gloves that suitably protect the hand.

- Using battery-powered tools is a good way of preventing electric shocks. Such tools also give mobility.

POINTS TO REMEMBER

Cover metal handles with plastic or tape to avoid electric shocks and to increase grip comfort. Purchase or choose tools with such handles.

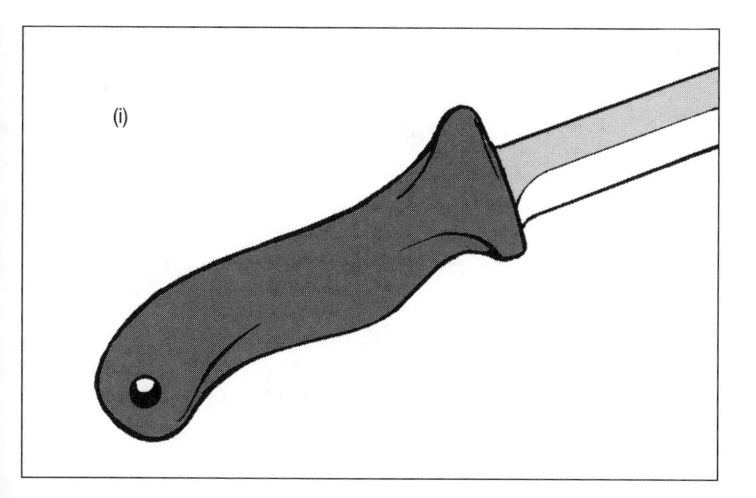

Figure 31. (i) and (ii) For metal handles, provide proper insulation that can prevent burns and electric shocks

CHECKPOINT 32

Minimize vibration and noise of hand tools.

WHY

Vibration from the tool transmitted to the hands not only disturbs the tool operation but injures nerves, tendons and blood vessels.

A hand-tool operator is always near the noise source. Noise damages hearing and hinders communication with other workers.

Exposure to hazardous vibration and noise is particularly significant for hand-tool operators because they are exposed for as long as they work.

HOW

1. Separate a noisy hand-tool operation from other parts of the workplace, for example, by means of partitions or placing the noisy operation in another small room, so that any noise affects only the tool operator. The inverse square law says that every doubling of distance reduces noise by 6 dB.

2. Purchase hand tools with low vibration and a low noise level. This requires making vibration and noise specifications part of the purchase orders and purchasing tools with good enclosure, vibration buffers and noise mufflers.

3. For pneumatic (air-powered) tools, use pressure regulators so that the tool operates at the design pressure instead of full line pressure.

4. Use automatic shut-off (i.e. turning off the noisy machine or tool when it is not in operation). This ensures the least possible exposure to vibration and noise, and saves energy.

5. Consult a noise specialist about how to reduce noise and vibration (noise specialists are also knowledgeable about vibration).

SOME MORE HINTS

- Electric hand tools are often quieter than pneumatic hand tools.

- Maintenance greatly helps to keep vibration and noise levels to a minimum. Tighten screws and bolts. Sharpen tools. Lubricate bearings. Grease and oil parts. Rebalance rotating equipment. Replace leaky compressed air valves.

- Orient workstations so that noise from the neighbouring workstation hits the ears of the others from the rear (best) or front rather than the side. This may reduce the noise effect by 5 dB.

- Provide good personal protective equipment against vibration and noise.

POINTS TO REMEMBER

Make noise and vibration specifications part of the purchase order of power tools. Also ensure automatic shut-off to minimize exposure.

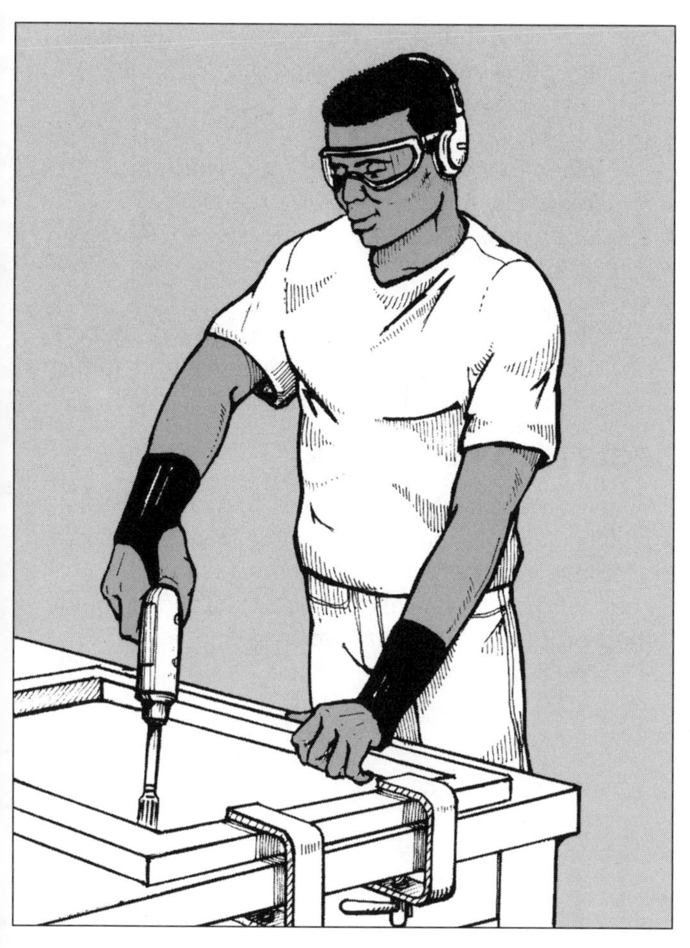

Figure 32. Provide good protection against noise and vibration

CHECKPOINT 33

Provide a "home" for each tool.

WHY

If each tool has a "home", i.e. a special and permanent place allocated to it, workers can find the tool quickly and are encouraged to use the right tool at all times.

If tools do not have "homes", some workers lose time searching for lost tools. Providing a "home" for each tool is an effective way to prevent this time loss.

Tools stored in their special places can be seen at a glance. Their inventory is therefore easy. This is a great help for good maintenance.

HOW

1. There are various means of providing a "home" for each tool. This can be a special shelf, a drawer, a particular place on a rack, an easy-to-see container, a tool trolley, a hook on the wall, suspension from an overhead structure, or a tool board. The most appropriate means should be chosen considering the size, shape and weight of the tool.

2. Do not forget to find a "home" also for bigger tools. Avoid the practice of putting large tools on the floor.

3. When various small tools are used, provide a tool storage board or special containers in which each tool has its own "home". A specially designed tool board for this purpose is useful.

4. In the case of a tool board, the outline of each tool could be drawn to show where it goes. Alternatively, labels could indicate where each tool goes.

5. The more frequently a tool is used, the closer its "home" should be to the worksite where it is used.

SOME MORE HINTS

- A series of small tools or tool parts of similar kind (such as tapes, drills, cutters, etc.) can be stored in special bins, trays or inserts with labels or a clear indication for each item. The necessary parts can then be taken out at a glance and put back easily.

- When the worker or a group of workers frequently change worksite, use portable toolboxes, tool trolleys or mobile tool storage racks.

- Suspended tools do not crowd the workbench and can be easily grasped. They always return automatically to their suspended "home".

POINTS TO REMEMBER

Without clearly designated "homes" for tools, it is difficult to put different tools in order. By providing a "home" for each tool, we can avoid time lost searching for tools. This is a good starting-point for their proper use and maintenance.

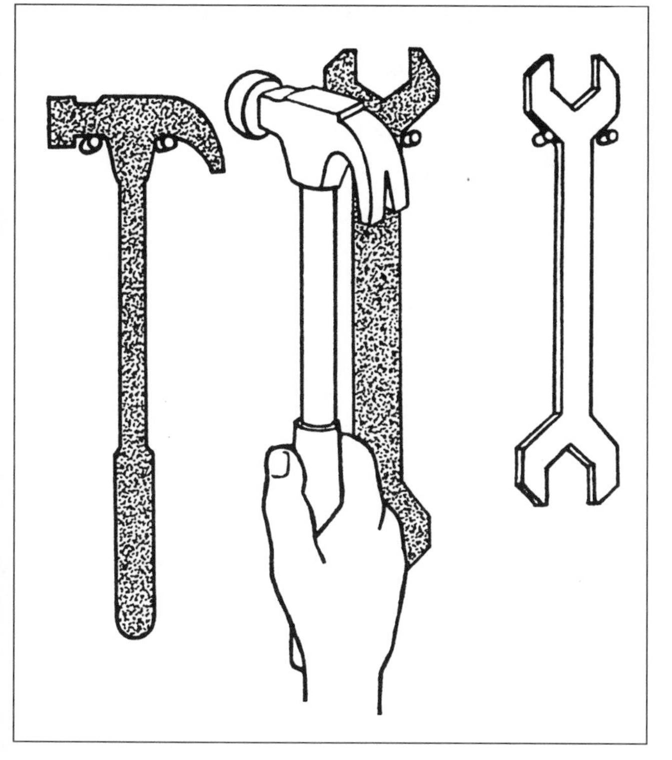

Figure 33a. The outline of each tool should be drawn on the tool board to show where it goes. This helps maintain order and immediately shows if anything is missing

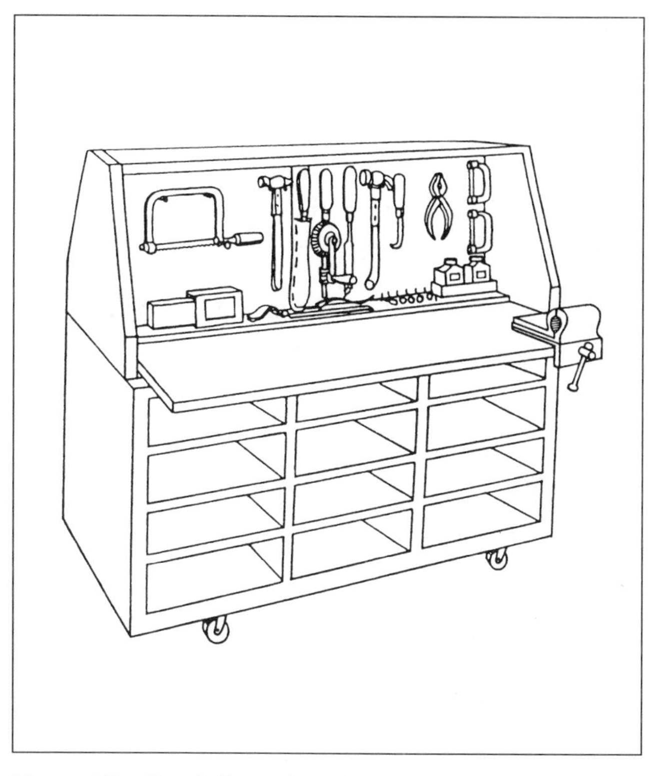

Figure 33b. A mobile workstation for a metalworker

Figure 33b. Provide "homes" close to the worker for tools used repeatedly. "Homes" for tools used less frequently can be placed around the workstation

CHECKPOINT 34

Inspect and maintain hand tools regularly.

WHY

Tools which do not work properly increase the operator's downtime and thus lower productivity.

Poorly maintained tools can cause accidents. The result may be serious injuries.

Regular maintenance of tools should be part of good management. Cooperation of all workers in this respect has positive effects on production and human relations.

HOW

1. Begin by purchasing reliable hand tools. Insist on always using reliable hand tools. Instruct all workers to replace a failed hand tool quickly.

2. Establish routines for regular inspection of hand tools. Some tools can be inspected by workers themselves, while others should be inspected by qualified personnel.

3. Provide a spare hand tool, or spare parts, on site.

4. If possible, provide substitute modules that can be used for replacing broken tool parts. Such substitute modules are easy to handle and permit quick repair by unskilled workers. All that is needed in the case of tool trouble is to take out the broken module and install a fresh one. Repair can be done later by the supplier or by skilled workers.

SOME MORE HINTS

- Maintenance time (the time taken to inspect and fix a tool) is usually small compared to the time taken to find that the tool does not work, to find the problem and (especially) to obtain the repair parts.

- Increased downtime (time taken to locate the fault, obtain parts and carry out repairs) means decreased worktime for the tool. Reduce this downtime by arranging replacement of parts/modules in advance.

POINTS TO REMEMBER

Dull tools require extra effort and reduce accuracy. Therefore consider "preventive maintenance", i.e. fixing things before they break. This is very important for maintaining tools.

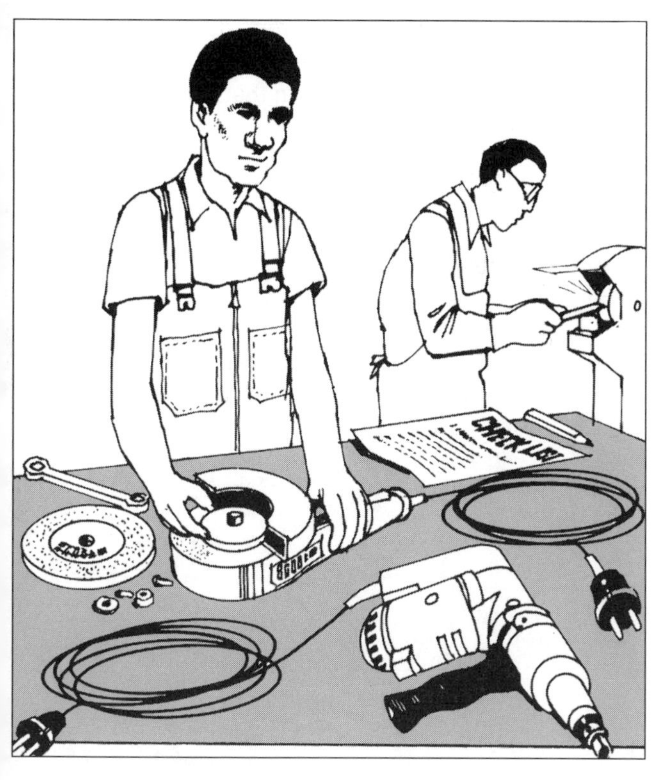

Figure 34a. Servicing tools is very important and repairs should be carried out by properly skilled tradespeople

Figure 34b. Provide adequate places for regular maintenance and repair of tools

CHECKPOINT 35

Train workers before allowing them to use power tools.

WHY

Power tools can increase production because they are faster and stronger than people. However, these advantages can be lost if they are used incorrectly.

Power tools are stronger than non-powered tools, and therefore accidents caused by their incorrect use will be more serious.

Power tools are always used for specialized tasks that require skills. Train and retrain workers for better skills and safety.

HOW

1. When purchasing power tools, make sure that they come with good instructions about their proper use.

2. Identify errors, scrap loss, injuries and slow cycle times caused by improper use of power tools. Interviewing workers gives you useful information, too.

3. Arrange for time to train and retrain those who use power tools in the correct operation of the tools.

4. Safety should always be an important part of such training.

5. Identify workers who are skilled at operating power tools and get them to train others in how to achieve the same high productivity and safety.

SOME MORE HINTS

- Consult instructions or manuals that come with the power tools for more hints on proper operation.

- Ask those who use power tools which tools are most difficult to operate. These problems may disappear with training.

- Training in tool use is an important part of new employee training. It is easier to train workers before they develop habits than to get them to break bad habits.

POINTS TO REMEMBER

Protect people and equipment by making sure workers are using their tools safely and productively. You must take time for training.

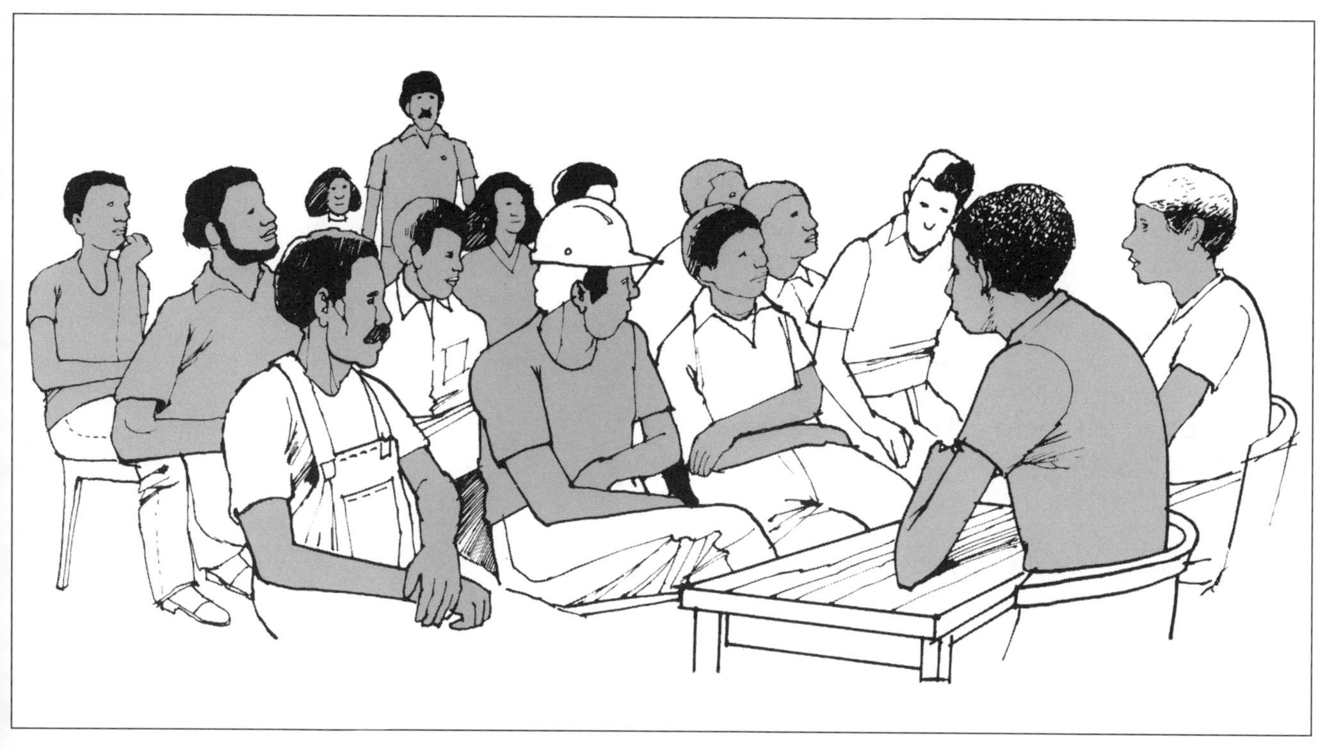

Figure 35a. New employees must be given full information by the supervisor and by co-workers. Information must also be provided when introducing new machines so that the work may be carried out as safely as possible

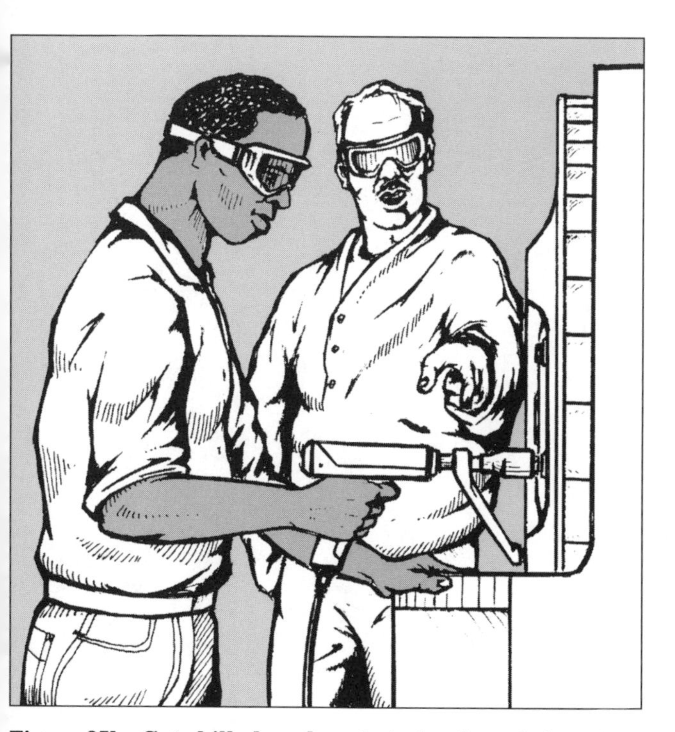

Figure 35b. Get skilled workers to train others in how to achieve the same productivity and safety

Figure 35c. Workshops should have their own safety rules displayed on the workshop wall. Learn these rules

CHECKPOINT 36

Provide for enough space and stable footing for power tool operation.

WHY

A stable posture should be used while operating a power tool. Proper footing is always necessary. Productivity and efficiency of power tool operation are thus significantly increased.

Loss of control during power tool operation is very dangerous and fatiguing. There must be enough space for the operation and for good footing.

HOW

1. Make sure that the floor for the power tool operation is flat, even and non-slippery. Provide an appropriate platform if necessary.

2. Remove potential obstacles to the tool operation.

3. Provide sufficient knee clearance, as well as foot clearance, so as to allow a stable posture close to the point of operation.

4. Provide emergency or automatic shut-off in case the worker slips.

SOME MORE HINTS

- Adjust the working height for each worker so that the power tool is operated somewhat below the elbow level and in front of the body while proper footing is ensured.

- Check whether the worker's footwear is suitable for the operation and for safety.

- If appropriate, consider the provision of a balancing, suspension or sliding mechanism or a support stand to allow for a stable posture during the tool operation.

POINTS TO REMEMBER

Ensure that the workplace allows a stable posture with proper footing while using power-driven tools.

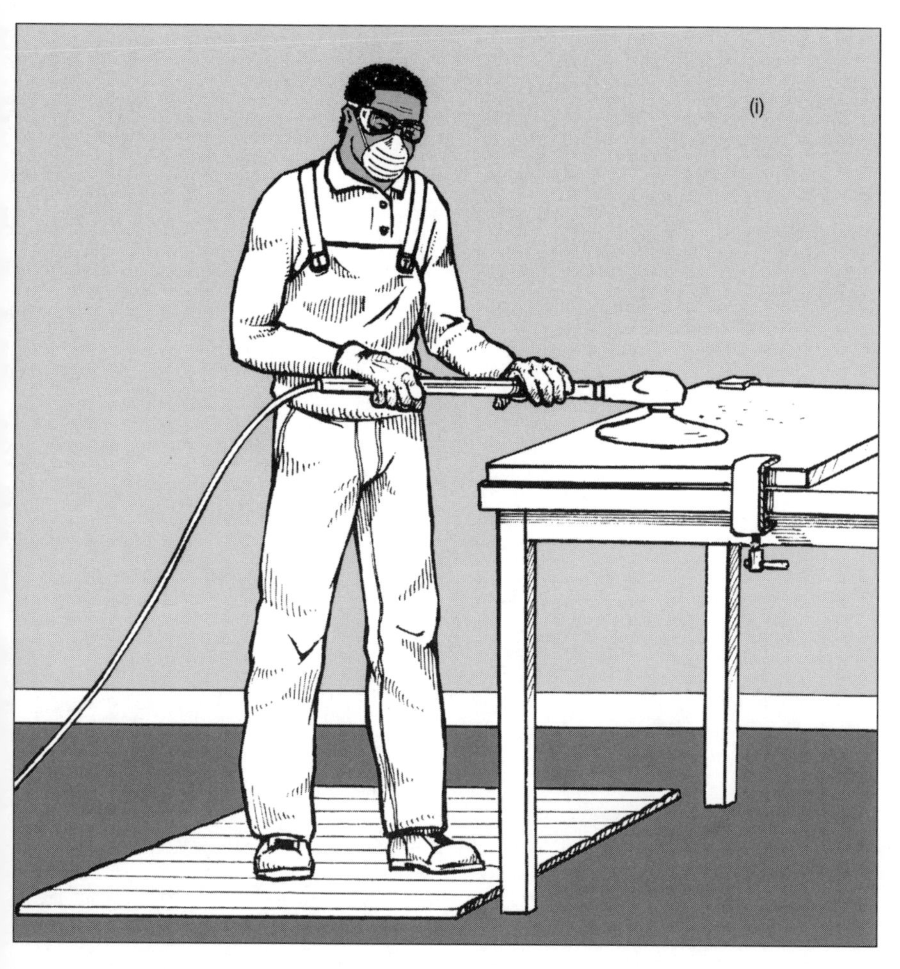

Figure 36. (i) and (ii) A stable posture with proper footing and appropriate work height around the elbow level is always necessary for safe and productive power tool operations

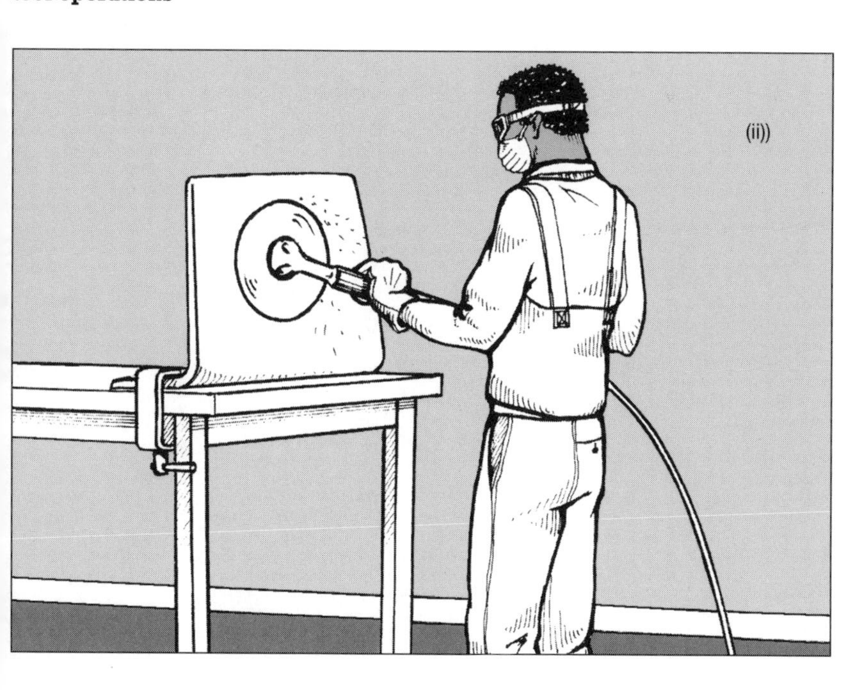

Productive machine safety

CHECKPOINT 37

Protect controls to prevent accidental activation.

WHY

Accidental turning on or off can cause serious injuries or damage and lower productivity.

Accidental activation can happen especially when many controls are situated in a narrow space.

Where accidental activation is prevented, the workers feel more secure and can concentrate properly on their tasks.

HOW

1. Cover or "cage" controls that are likely to be turned on or off accidentally.

2. Choose a control that works in the opposite way that an accidental activation is likely to occur. For example, if people are likely to activate a control accidentally by leaning on it and pressing, then choose a control that requires pulling to activate it. But consider that controls should still be moved in the expected direction.

3. Replace existing controls that can be accidentally activated with controls that have more resistance and are harder to activate. However, the controls should not be so difficult to use that operators cannot activate the control when they want to do so.

4. Position particularly important controls, such as power on-off or emergency switches, away from other controls. This helps avoid inadvertent activation during normal operations. Naturally, controls should still be within easy reach.

SOME MORE HINTS

- Make sure covers and cages do not hide the control or confuse the worker. If the control protector makes the control difficult to see, consider using a clear or see-through device.

- When purchasing new equipment, look for machine designs that minimize accidental activation. Useful designs include: mounded controls; recessed controls; controls that require two different actions (e.g. pull towards the operator and then pull towards the floor); or two controls required to activate.

POINTS TO REMEMBER

Turning a machine on or off unnecessarily is dangerous to people and bad for equipment, and slows production. There are different ways to prevent this unnecessary activation.

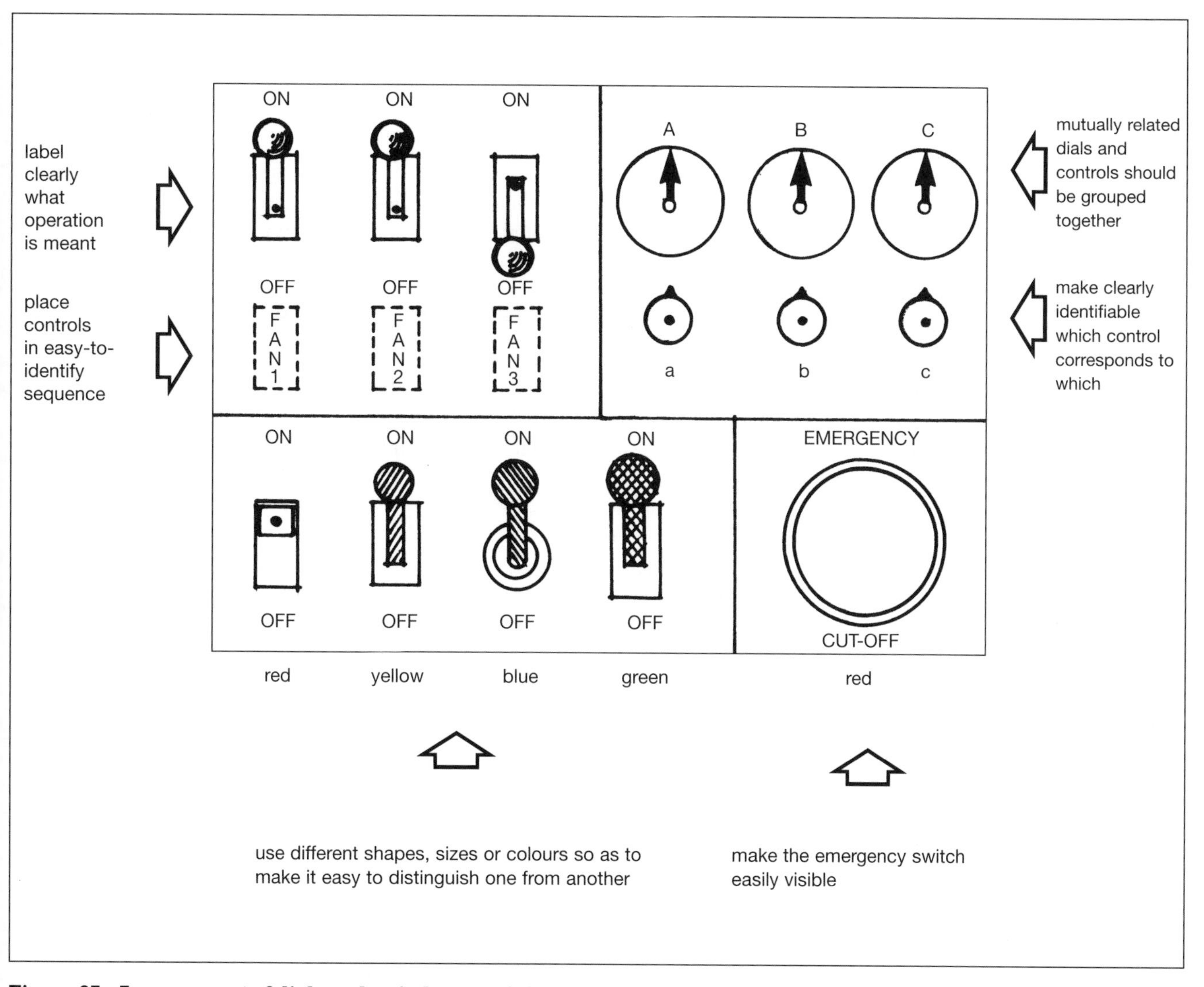

Figure 37. Arrangement of dials and switches to minimize mistakes

CHECKPOINT 38

Make emergency controls clearly visible and easily accessible from the natural position of the operator.

WHY

Emergency situations are stressful, and operators are likely to make mistakes. Emergency controls must therefore be particularly well designed so that fast action is possible without any mistakes.

In an emergency, it may happen that the operator in charge is absent or injured. Co-workers who are trained in advance about emergency operation may have to act quickly. It is essential to make emergency controls easy to find.

Even untrained co-workers must be able to find emergency controls.

HOW

1. Make emergency controls or cords easy to reach. Put them in a location that is natural for the worker to reach (e.g. not by twisting the body).

2. Make emergency controls large and easy to activate. For example, use a large rather than a small push-button.

3. Colour emergency controls red.

4 Check to make sure that these controls are in line with regulatory standards.

5. Position emergency controls away from other frequently used controls, thereby reducing the risk of inadvertent activation.

SOME MORE HINTS

- Many types of emergency control are used. In addition to palm buttons and emergency cords, a dead man's switch may be used. As long as the switch is actively pressed the machinery keeps going. If the pressure is released the machinery stops.

- Make it possible for the machine to switch itself off automatically in the event of a worker inadvertently coming into a danger area. For example, some rotating machinery has emergency trip-cords located above the operator's feet; if the operator is pulled into the machine, the feet will catch the trip cord and the machine will stop.

- Think of innovative ways to automate emergency action. For example, a worker could step on a "pressure mat".

POINTS TO REMEMBER

Emergency situations are very stressful. Even trained workers may make mistakes. Emergency controls must be designed so that there is no chance of mistakes in activating the controls.

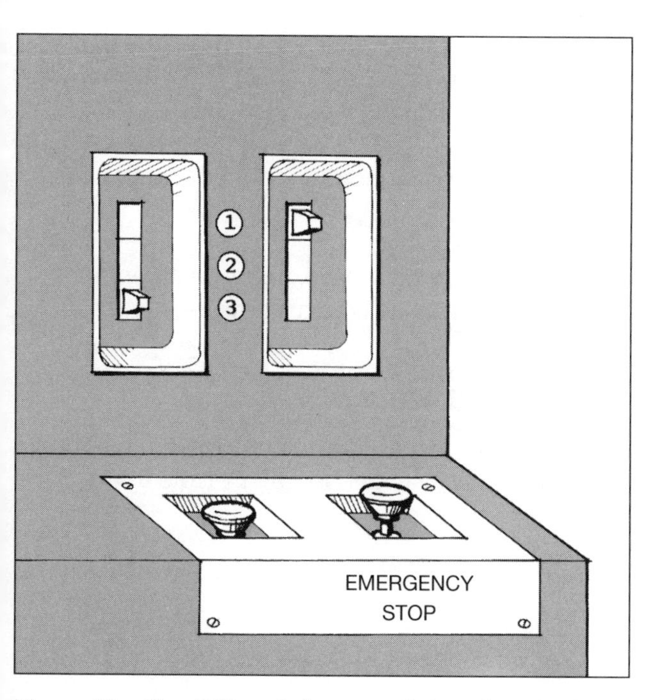

Figure 38a. Use different shapes and sizes to make it easy to distinguish one control from another

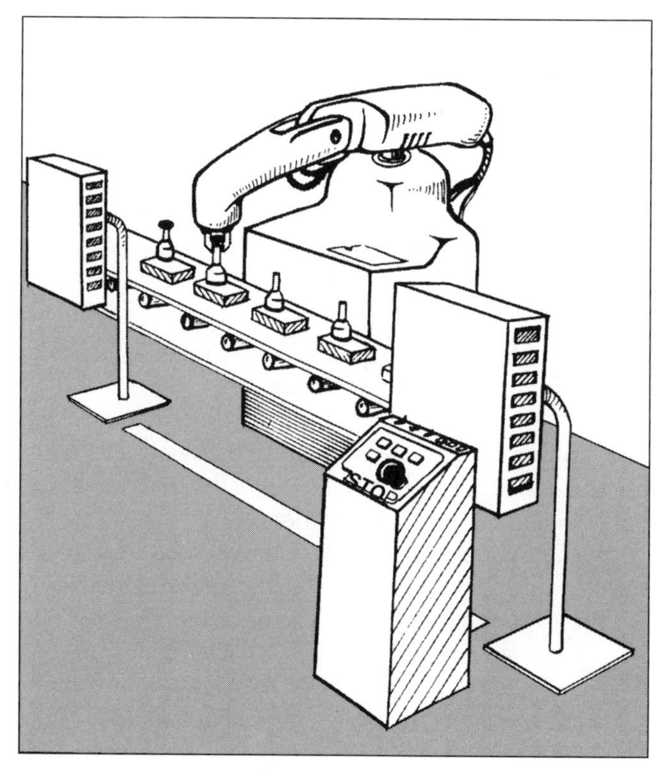

Figure 38b. Make emergency switches easily visible

CHECKPOINT 39

Make different controls easy to distinguish from each other.

WHY

If controls look similar, people will make mistakes. Activation of a wrong control may lead to an accident.

Controls that are quick and easy to find will save time and reduce operator errors.

Controls are sometimes easy to distinguish just because they have different locations. But often this is not sufficient. By adding another feature, such as colour, size, shape or labels, controls are much easier to distinguish from each other. This is called "coding" of controls.

HOW

1. Use different colours, sizes or shapes for switches and other controls:

   - use different colours for different controls;
   - use controls of different sizes;
   - use control knobs of different shapes.

2. Label the controls. Attach clearly visible, simply worded labels. Use labels written in the local language.

3. Standardize the location of common controls on similar machines. For example, place controls in an easy-to-identify sequence (from fan 1 to fan 2 to fan 3, etc.) or in a place where it is easy to identify which control corresponds to which display (placing the heat-controlling knob directly under the temperature dial, etc.). In this way, the control panels for similar machines should also look alike. This will reduce errors in operation.

SOME MORE HINTS

- Make emergency controls (such as an emergency cut-off switch) look very different and easily visible by means of colour, size and shape.

- Use no more than three different sizes of control knob, because people cannot distinguish more than three sizes.

- The shape of a control knob can be made to look like the controlled function (e.g. a control for a fan can look like a fan).

- Colour coding does not work in dark environments.

- Labels can be put above, underneath or at the side of controls, as long as they are clearly visible.

POINTS TO REMEMBER

Coding of controls (by colour, size, shape, label and location) can prevent operator errors and reduce the time for operation.

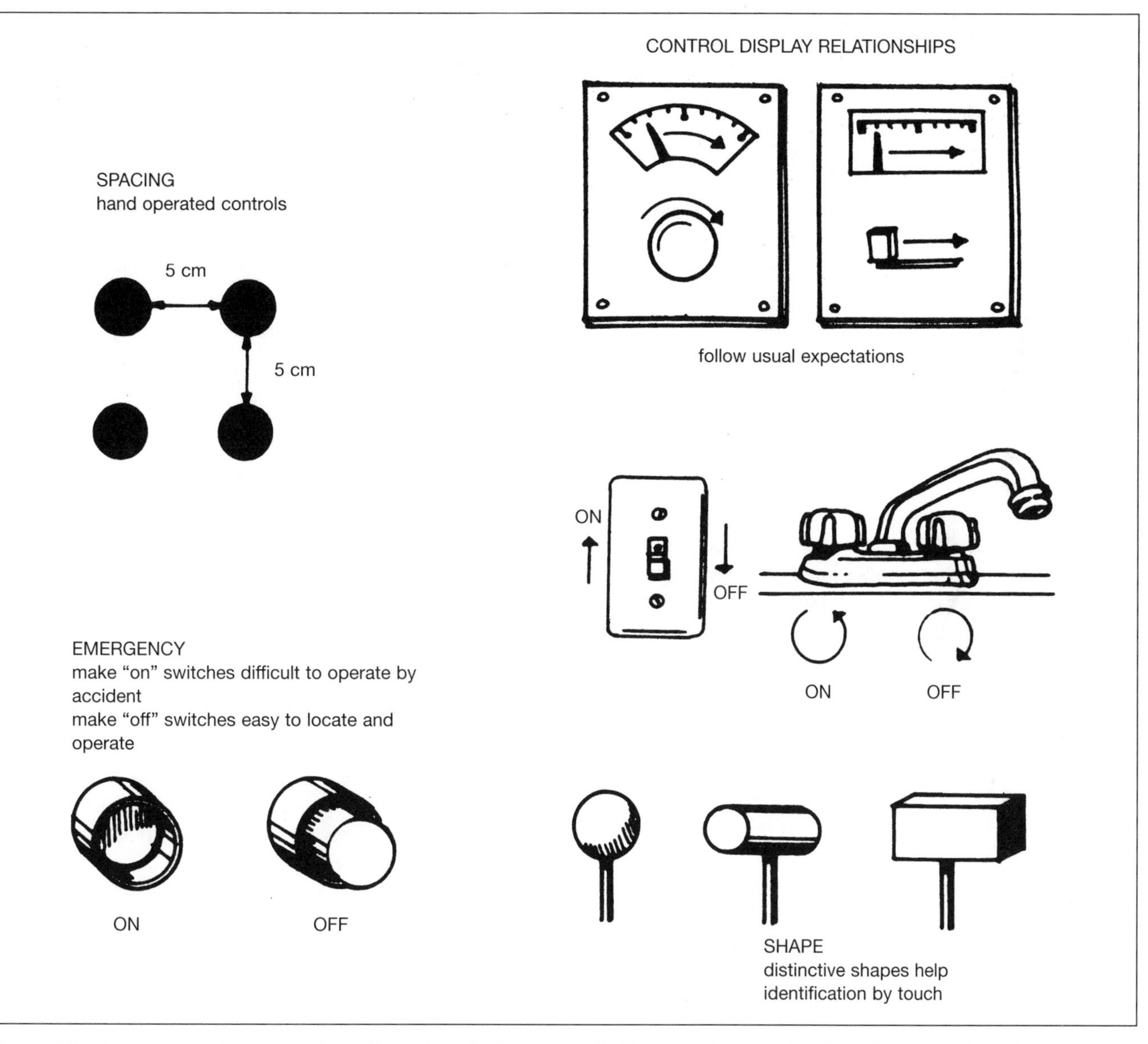

Figure 39. Try various ideas to make different controls easy to distinguish from each other. Grouping, keeping good control-display relationships, spacing, different shapes and different colours are all useful ideas. Making emergency "off" switches easy to distinguish is particularly important

CHECKPOINT 40

Make sure that the worker can see and reach all controls comfortably.

WHY

All items that are touched by the hands need to be organized. These include controls, hand tools, parts for assembly and part bins. In many cases workers themselves organize these items at the workstation, but often they do not.

If controls are not easy to see or reach, operators tend to use them by relying on habits and guess-work. This can cause mistakes.

Time and effort are saved by placing controls within easy reach. Controls placed too high cause shoulder pain and controls placed too low cause back pain. It is important to locate them in a place that is easy to reach from a normal working posture.

HOW

1. Place the most important controls (primary controls) in front of the operator so that the control operation is done at around elbow height without bending or twisting the body.

2. Controls of secondary importance may be placed next to the most important controls. In any case, they should be within easy reach from the normal working position. Avoid places where twisting of the body becomes necessary for operating the controls.

3. If control positions are too high, use a platform to raise the floor on which the worker stands or sits for work. If control positions are too low, try to raise them by relocating them or putting a platform under the machine or workbench.

4. When new workstations or new machines are introduced, purchase those which are suited to the size of the operators or those in which the workstation or control heights are adjustable. Most work operations are best done around elbow level. This "elbow rule" can be applied to determine the correct hand height during operations.

SOME MORE HINTS

- It is useful to identify the primary hand movement area (between 15 and 40 cm from the front of the body and within 40 cm from the side of the body at elbow height) and the secondary hand movement area (beyond the primary area but within 60 cm from the side of the body at elbow height). Position primary controls and other primary items (hand tools, parts) within the primary area, and secondary controls and other secondary items within the secondary area.

- See that the controls are located in a good combination with other items, such as tools, parts to be grasped, semi-products to be placed on the workstation, bins, etc. Try to organize the layout of all these items based on the opinion of the experienced workers.

- The work-table surface may sometimes be divided into subtask areas where operations are done sequentially. This helps to organize the task and facilitates learning and productivity.

POINTS TO REMEMBER

A well-organized workstation will save time and is productive. The location of controls according to their primary and secondary importance helps to organize workstations.

Figure 40a. Ensure that the worker can reach all controls comfortably from a standing or sitting position

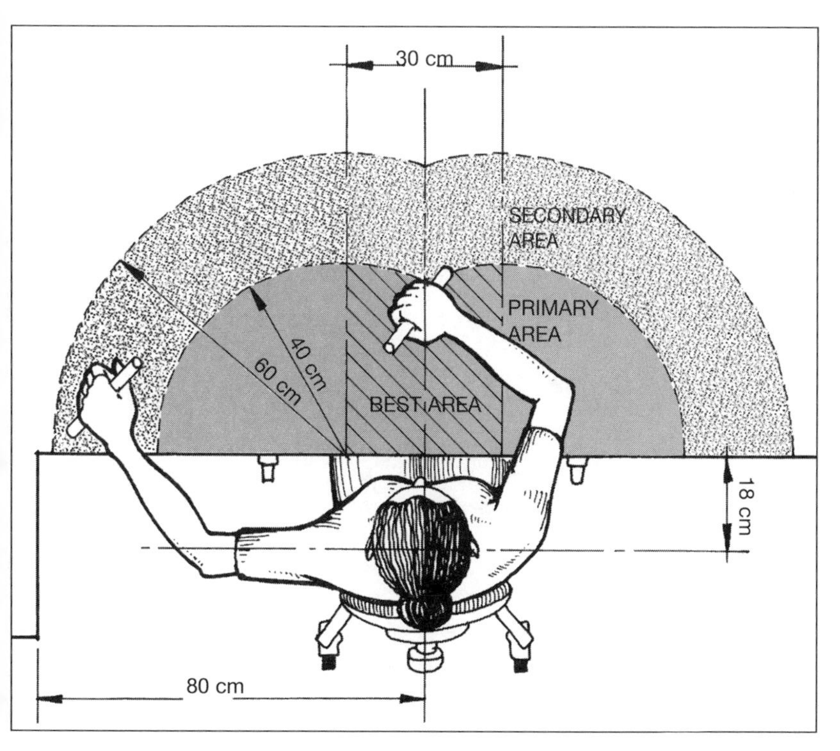

Figure 40b. All controls need to be within easy reach of the worker and easy to see

CHECKPOINT 41

Locate controls in sequence of operation.

WHY

Some machines have multiple controls that are difficult to learn to operate. For example, hydraulic equipment used in mining or manufacturing may have 10-12 controls. The sequence of control operations can be made easy to learn if controls are positioned to follow the task.

When multiple controls correspond to multiple machines or machine parts, controls are often confused. This can be avoided by locating controls in the same sequence as they are operated.

By placing controls in a logical sequence, it is easier to standardize their locations between similar machines. This greatly facilitates learning.

HOW

1. Identify subtasks in machine operation, such as "power on-off", "setting up", "operating" or "moving machine". See if controls for each subtask are easy to distinguish from those for other tasks.

2. Reposition controls by changing electrical connections (or even changing hydraulic hoses, if possible) so that controls associated with each subtask are grouped together.

3. Position controls according to the sequence of operation within each subtask (e.g. position controls A, B and C in this sequence when the corresponding operations A, B and C are done in the same sequence).

4. Similarly, position controls according to the different machines or machine parts (e.g. position controls A, B and C in the same sequence as the corresponding machines A, B and C).

5. Code the controls by colour, size, shape or label to make it easier to distinguish between them.

SOME MORE HINTS

- Workers sometimes modify controls or control knobs to make them easier to operate. Look out for such modifications, since they indicate that there is a need for change.

- Make a list of the different subtasks and the sequence of control operations. Ask the workers to help out and verify this information. Then consider if changes in location will be useful.

- Ask the workers if controls corresponding to different operations are easy to find without mistakes. If not, try to change their locations or introduce coding.

POINTS TO REMEMBER

Workers will make fewer errors if controls are located in an easy-to-understand sequence. This improves both safety and productivity.

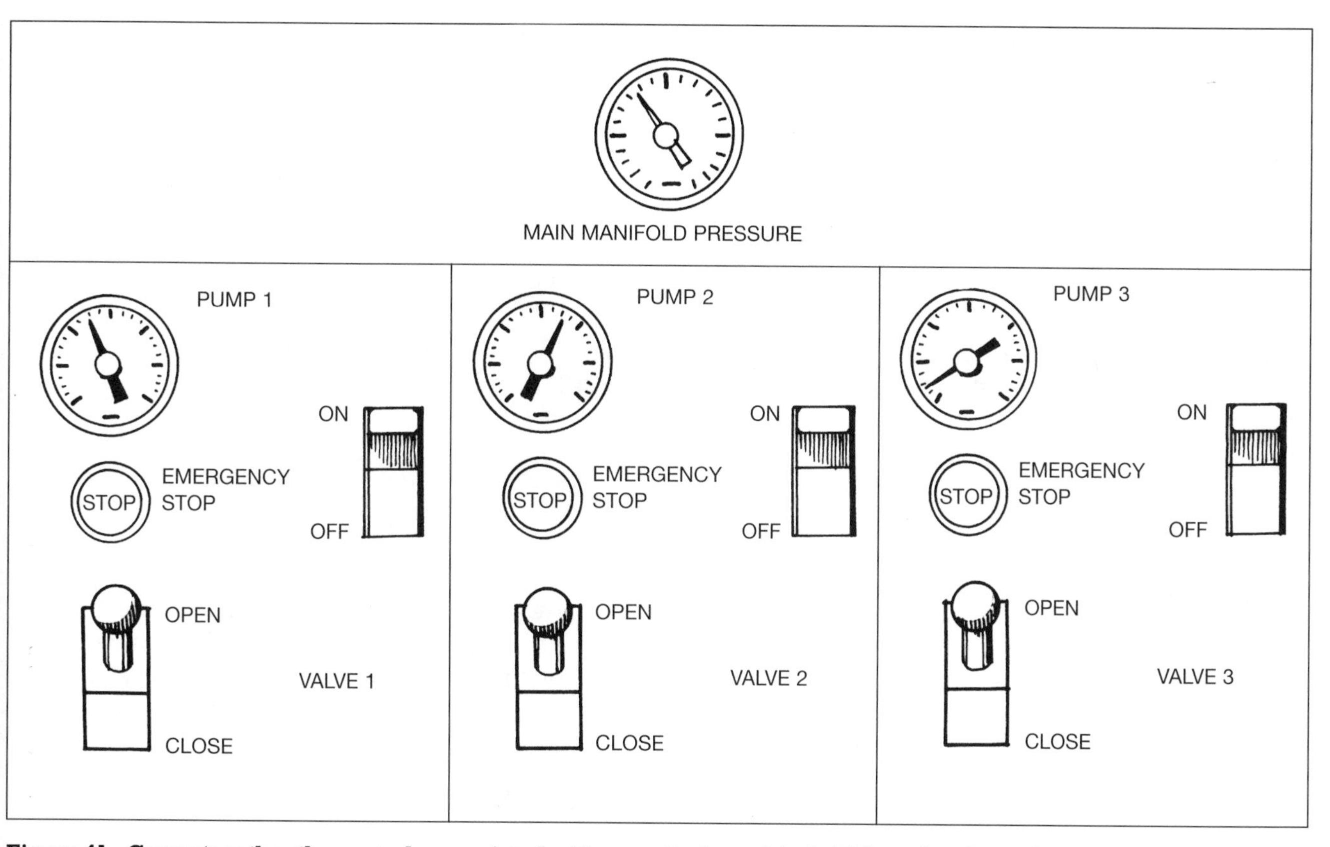

Figure 41. Group together the controls associated with a particular subtask. This makes it much easier to locate them

CHECKPOINT 42

Use natural expectations for control movements.

WHY

Most people have expectations of how a control should be moved.

In a car, there is a clear expectation to move the steering-wheel in the same direction as the road turns. A car designed differently would be a disaster. The same principles apply to machine controls.

Note that expectations may be different in different countries. For example, in many countries (e.g. India) a light switch is turned down to turn on the light, while in others (e.g. the United States) the light switch is turned up.

HOW

1. Use the expectations in the following table.

Desired action	*Expected control movement*
Turn something on	Right or forward or clockwise or down (up in some countries)
Turn something off	Left or back or anti-clockwise or up (down in some countries)
Move something right	Right or clockwise
Move something left	Left or anti-clockwise
Raise something	Up, back
Lower something	Down, forward
Retract something	Pull back or up
Extend something	Push forward or down
Increase something	Up or right or clockwise
Decrease something	Down or left or anti-clockwise
Open a valve	Anti-clockwise
Close a valve	Clockwise

2. Make sure that the control movements of different machines or power switches use the same principles.

SOME MORE HINTS

- Some control expectations are "more natural" than others. For example, to raise an overhead crane, a horizontal control would move up, but a vertical control would move backwards. For the horizontal control there is a one-to-one correspondence between the movement of the control and the crane. This is a strong expectation.

- For a vertical control that pulls back and forth, expectations would be more mixed, since there is no clear one-to-one correspondence. A few people would probably push the control forward to raise the crane. It is better to avoid this kind of confusing control movement.

- Keep the dial movement and control movement corresponding with each other. For example, if the dial pointer moves to the right by increasing something, the control placed beneath should also be moved to the right (or clockwise) to increase it.

POINTS TO REMEMBER

People have expectations of how to move controls. Do not violate them. Use these expectations to your advantage to reduce control errors and to make production more effective.

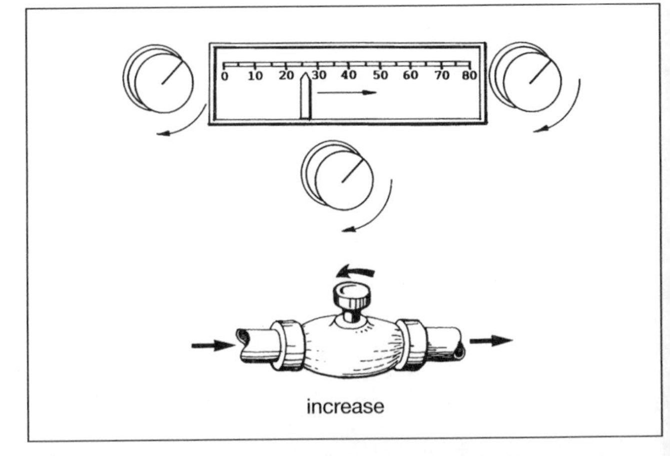

Figure 42a. There are certain established relationships between control movements and the decrease-increase effects. Ensure that they are in accordance with the local workers' understanding

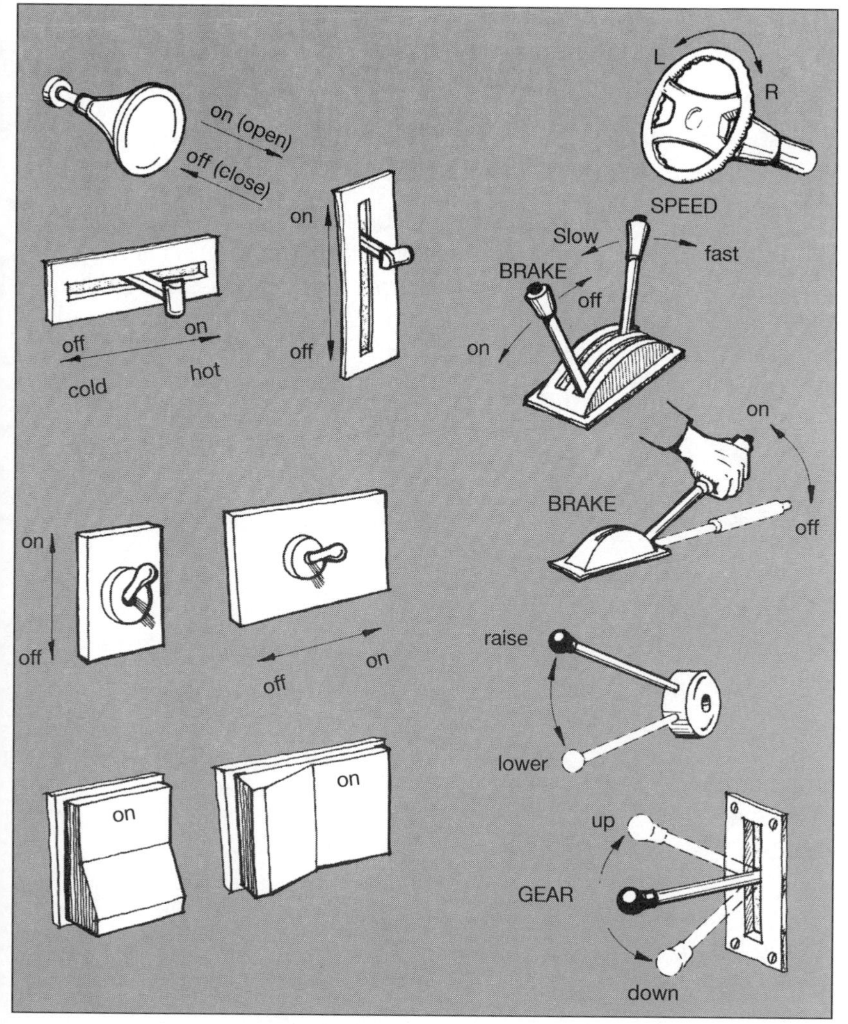

Figure 42b. Use natural expectations for control movements. Be aware that these expectations may differ between different populations. For on-off movements, follow the local habits, but with clear "on" and "off" signs. For right-left, up-down and other movements, make sure that workers fully understand the correct directions

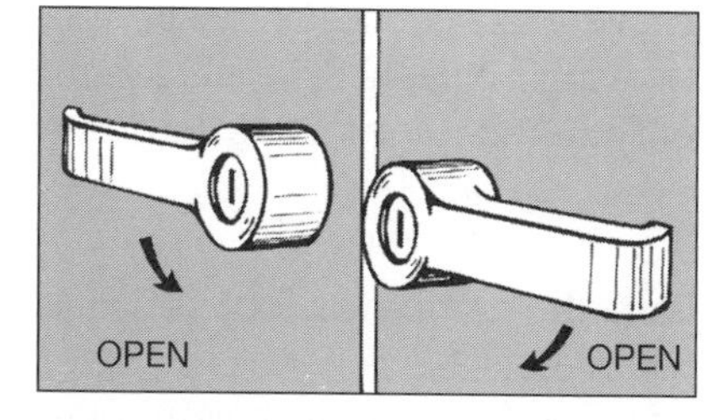

Figure 42c. The movements of door controls relate to how people understand the control effects. Adding "close" or "open" signs can always help

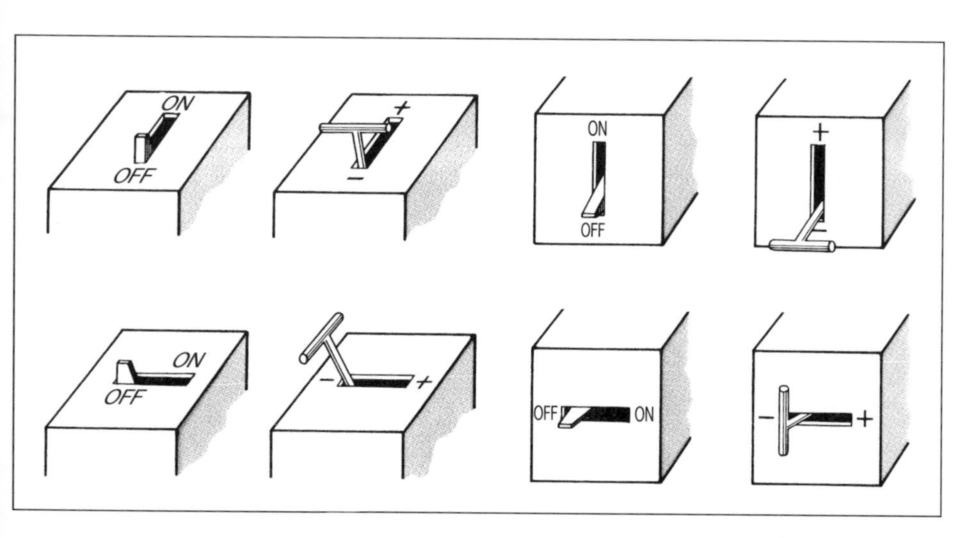

Figure 42d. Direction of controls which are easy to understand for most people. Adding signs is always helpful

CHECKPOINT 43

Limit the number of foot pedals and, if used, make them easy to operate.

WHY

Foot pedals can be useful as alternatives to hand controls. One of their major benefits is when both hands are busy. The use of foot controls can also free space in a workstation. Foot controls, however, often require keeping a special posture and thus restrict the operator's movement. This is particularly critical for standing operators.

Foot pedals that are operated repetitively by one foot cause one-sided strain which may lead to back pain.

Foot pedals cannot be easily seen from the normal working position. Special care must be taken to prevent stumbling or inadvertent activation.

HOW

1. Limit the number of foot pedals to a minimum when their use is necessary. Avoid as much as possible foot pedals that are operated repetitively by one foot only.

2. Locate a foot pedal at floor level in order to avoid uncomfortable foot positions. A pedal level that can be reached only by raising the foot away from the floor is uncomfortable and forces the worker to maintain an unnatural posture.

3. Make it possible to move the location of a foot pedal on the floor.

4. Make the foot pedal large enough to fit the sole of the foot.

5. Consider using a foot-rest at the side of the pedal.

SOME MORE HINTS

- Foot controls are good for many applications if adequate care is taken about the working posture and ease of operation. They are even used as a cursor control for computers (a "foot mouse").

- Be careful in locating foot controls, since a foot control may be a tripping and falling hazard.

- Adjustability of the location of foot pedals is important to improve workers' comfort and convenience. This is helpful especially for standing operators.

POINTS TO REMEMBER

Foot controls are beneficial when the hands are busy with other tasks, and where there is limited space at the workstation. Make it possible to adjust the location of a foot pedal on the floor, particularly for standing operators.

Figure 43a. (i) and (ii) Locate a foot pedal at floor level and make the pedal large enough for easy operation

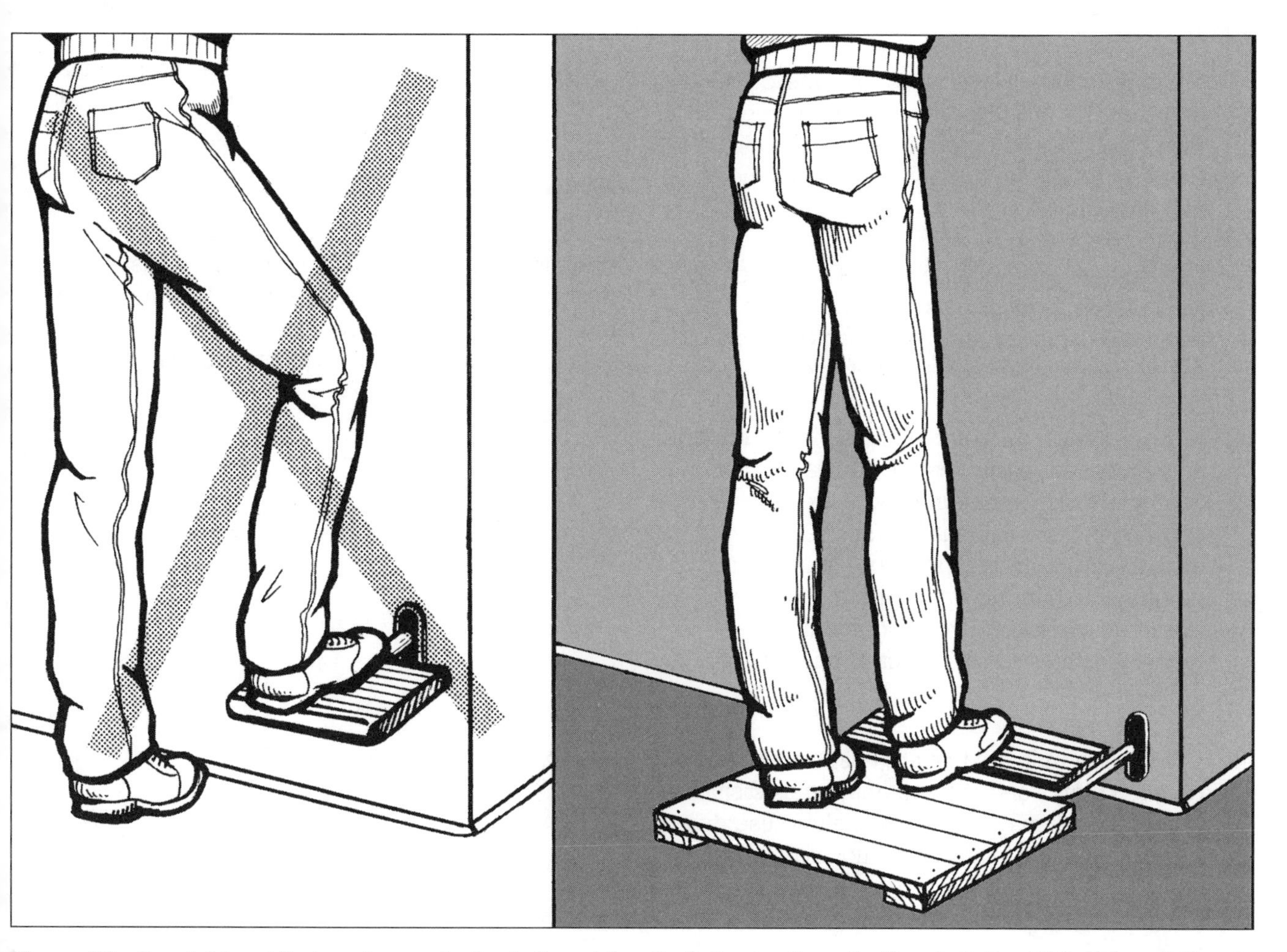

Figure 43b. A pedal level that can be reached only by raising the foot away from the floor is tiring. Make the pedal level lower and provide a foot platform for easy pedal operation

CHECKPOINT 44

Make displays and signals easy to distinguish from each other and easy to read.

WHY

Displays and signals carry information about work, and they should be easy to identify. It is important to consider the location of displays and signals, and also to make them easily distinguishable from each other.

Good visibility of pointer positions, characters and numbers on displays or signals also assures high-quality work.

Incorrect reading of displays or signals is sometimes critical, as this may lead to operation failures and accidents.

HOW

1. Put important displays or signals where operators are normally looking. Locate the most important ones at a viewing angle of about 20-50 degrees below the horizontal line from the operator's eyes.

2. Use different sizes, shapes or colours when different displays or signals are used by the same operator. Using colours for coding different information is often the easiest way to do this.

3. Make the characters and numbers large enough so they can be easily read at a distance. For example, for an operator viewing a display at 1 m distance in good illumination (say, 500-800 lux, as in the case of a well-lit office), a character height of 5-10 mm is appropriate. As viewing distances increase or reading conditions get worse, the size of characters should be increased.

4. Use display markers that are easy to read. Very detailed marks and crowded numerals disturb the reading. It may sometimes help to use different colours for different sections of a display.

SOME MORE HINTS

- Displays located in the peripheral viewing field are difficult to monitor. For example, if a display is located at more than 50 degrees from the central viewing point, the operator must turn the head to read the display. Under such conditions operators make more errors or even omit reading the displays.

- Good location of the displays, controls and corresponding machines is important. Use a layout for displays so that it is easy to understand the relation to machines and controls. It is very useful to group related displays and place them in sequence of operation. For example, displays placed just above the corresponding controls greatly help the operator in finding the control.

- Ensure good lighting on displays and signals in the evening or night hours.

- Displays can often be shown by means of a visual display unit (VDU). Presentation on a VDU screen offers an extra challenge, since the screen is small. Displaying information that is easy to understand and easy to read is equally important on VDU screens.

POINTS TO REMEMBER

Displays should be put in a location where operators look. Make different displays easy to distinguish from each other. Characters and scale markings must be of adequate size and clearly visible from the normal position of the operator.

Figure 44a. Locate the most important displays and signals at a viewing angle of about 20-50 degrees below the horizontal line from the eyes of the operator standing or sitting naturally

Figure 44b. If appropriate, select the pointers' normal positions so that they are easy to identify

CHECKPOINT 45

Use markings or colours on displays to help workers understand what to do.

WHY

For some tasks it may be necessary to display an exact numerical value, such as time in minutes. For other tasks it is enough to know if the value is within a certain range. One example is water temperature. It may be enough to know that the temperature is below boiling point.

Displays are there to help a worker carry out the right operation. Often workers themselves add markings to displays. Use these ideas to change displays to "helpful displays".

HOW

1. Add markings to indicate the point or ranges where a certain action is always necessary (e.g. temperature or speed-limit).

2. Use colour coding. For example, green areas or numbers mean acceptable. Red means unacceptable.

3. Group related displays together and organize them for easy inspection. For example, a break in the pattern of pointer positions is easy to see (e.g. if horizontal or vertical positions of all the pointers in the group mean that the operation is progressing correctly, then it is easy to find a pointer deviating from this pattern).

SOME MORE HINTS

- Usually two different types of display are available: (i) a counter with numbers; and (ii) a moving-pointer display which shows an approximate value. A moving pointer is appropriate for showing trends and changes (such as increases or decreases). In this case, the operator is not interested in detailed numbers.

- Position important displays where operators will normally be looking.

- Avoid parallax effects that occur when the position of a moving pointer somewhat above the dial surface is wrongly read by an operator looking at the dial from the side. Place the dial surface vertical to the line of vision or place such important displays in front of the operator.

POINTS TO REMEMBER

Displays should tell workers what to do. Use markings or colours for this purpose.

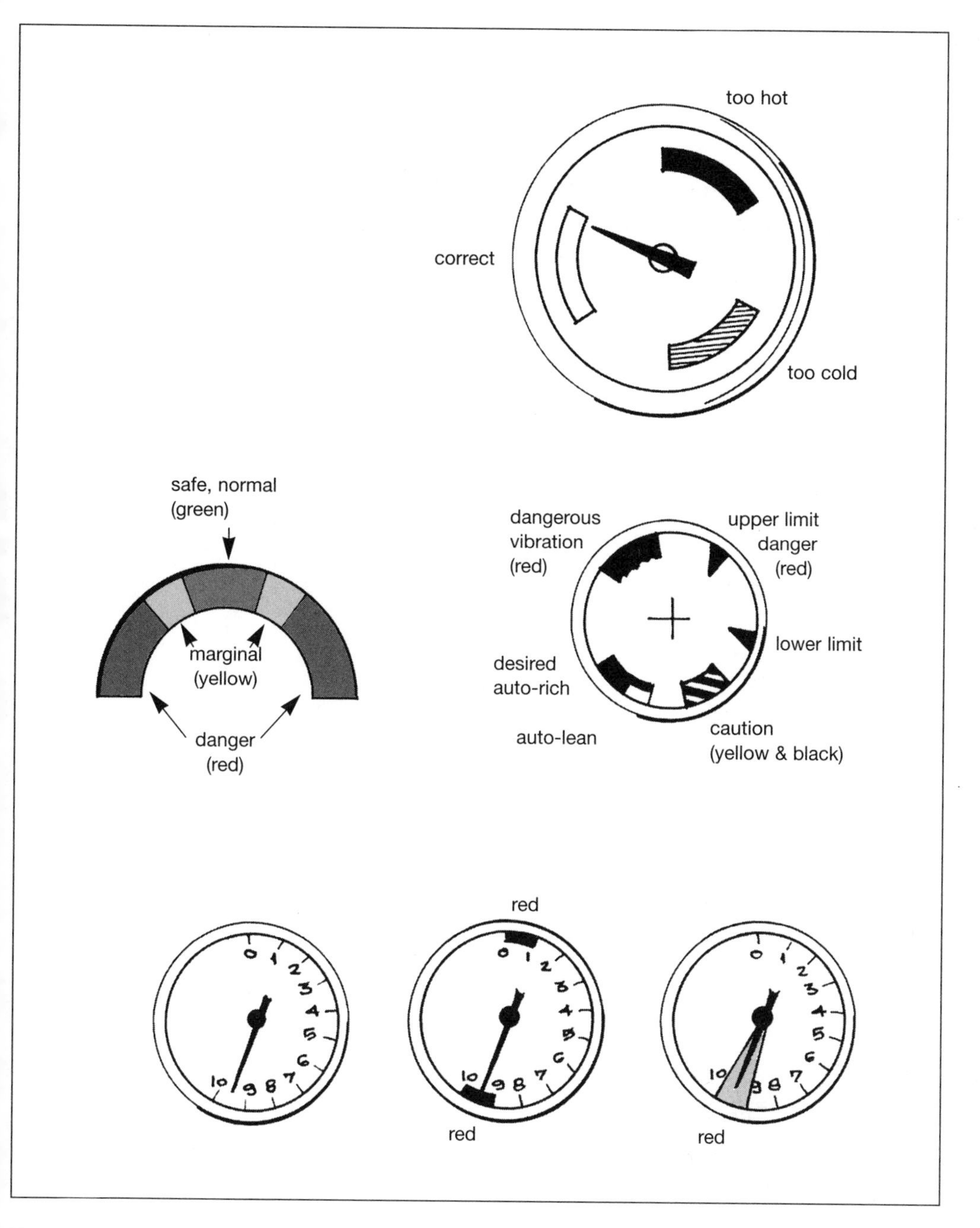

Figure 45. A display instrument should convey the required information as simply and unmistakably as possible

CHECKPOINT 46

Remove or cover all unused displays.

WHY

Often there are displays which are not used in operation. They continue to display unnecessary information simply because these displays were already there when the machine was purchased. Such displays are superfluous.

In critical situations essential information must be identified rapidly. Unused or superfluous displays may divert attention from essential displays.

Unused displays tend to be poorly maintained. This may create the impression that some equipment need not be well maintained, and may thus have an adverse affect on overall maintenance performance.

HOW

1. If there are displays which are clearly superfluous and not used, remove them from the display panel.

2. If it is difficult to remove the unused displays, hide them with adequate covers or paint.

3. When there is a chance to reorganize the display panel or install new equipment, make sure that only necessary displays are presented. Quite often, well-organized displays mean fewer displays.

SOME MORE HINTS

- Look for poorly maintained displays. There may be reasons for the poor maintenance, and one of them may well be the fact that these displays are not useful or not used at all. Seek the opinion of the workers concerned. If necessary, you can have a trial period by covering the unused displays with adhesive tape.

- The same can be said of signals or other forms of information which are superfluous and not used. After consulting the workers, remove them in order to make necessary information better understood.

- Modern complex machinery and automated systems use visual display units (VDUs) to present the needed information. However, following technical modifications, VDUs may also contain unnecessary information, which should be eliminated through reprogramming.

POINTS TO REMEMBER

In modern workplaces there are many display panels. Too many displays divert attention. Help workers by removing or covering unused displays.

Figure 46. There are many instrument panels in working life today where different types of signals, lights and gauges enable different features of the production process to be read rapidly. They should be arranged on a panel so that one look at the panel can show whether any of the gauges is giving an abnormal reading

CHECKPOINT 47

Use symbols only if they are easily understood by local people.

WHY

Symbols are sometimes used to identify machines, chemicals, controls and displays. In fact, many international manufacturers of machines prefer to use symbols, since they do not have to translate a label into the local language. But many symbols are difficult to understand, particularly those referring to machine functions that are hard to visualize or imagine. It is often better to use a short message instead.

Good symbols can be used in so far as they are easily understood by local people.

HOW

1. Use symbols only if you are absolutely sure that they are easily understood by all the workers concerned.

2. Simple symbols are better, but be aware that there are not many symbols that are universally understood.

3. Take several workers, one at a time, to the machine and ask them to identify the symbols. If a symbol is understood by all workers, there is no problem. If it is not understood by some workers, make a label and attach it to the machine.

4. Do not hesitate to add labels. They will prove essential in critical situations. The labels should be made to withstand wear and tear. Metal or plastic plates are the best solution.

SOME MORE HINTS

- Well-understood symbols have an advantage in that they are quicker to read than a label. There are widely accepted and widely used symbols, as in the case of no smoking signs, emergency exits and hazardous chemicals.

- If you want to propose your own symbols, get workers to evaluate them.

POINTS TO REMEMBER

Symbols that are difficult to understand should be replaced by labels. If in doubt, ask the workers.

Figure 47. Use symbols that are easily understood by local people

CHECKPOINT 48

Make labels and signs easy to see, easy to read and easy to understand.

WHY

Labels and signs must be easy to read, otherwise they will be ignored.

People tend to read labels and signs only at a short glance, and therefore often make mistakes in reading them. This may lead to wrong operation and may cause an accident. Labels and signs must be large and clear enough to be easily read at a distance.

Text must be made easy to understand so that people will know what to do. This is productive since it will save time.

HOW

1. Locate labels and signs in places where people often look, for example, close to the production process or in front of each operator.

2. In a workplace where the operator stays in the same place, locate labels and signs at a comfortable viewing angle from that position, i.e. about 20-40 degrees below the horizontal.

3. Make the lettering large enough to be easily read at a distance.

4. Where appropriate, use different colours or shapes for different labels or signals.

5. Put labels for displays and controls immediately above, underneath or to the side so that it is clear which label corresponds to which display or control. Make sure that these labels are not obscured by other elements.

6. Make the message clear and short. Avoid confusing and lengthy text.

7. Make sure that labels and signs use the language that can be understood by the workers. Where there is more than one language group, it may be necessary to use different languages in labels and signs.

SOME MORE HINTS

- Locate labels and signs so they do not pick up reflections from light sources which can cause glare. Sometimes you can change the angle of a sign to reduce reflections (as for a rear-view mirror in a car).

- Use materials such as plastic or steel that can easily be cleaned of dirt and oil, and so that the sign will remain visible for years to come.

- Labels with 1 cm high lettering are sufficient at workstations.

- When indicating a required operation, start the message with an action verb so that people know exactly what to do (e.g. "Turn off lights", "Hook the sling", not "Turn off lights if not necessary" or "Danger — Watch the crane").

POINTS TO REMEMBER

Labels and signs can give much important information. Locate them where the workers look, make them large enough and make the message short and easy to understand. This will reduce errors and save time.

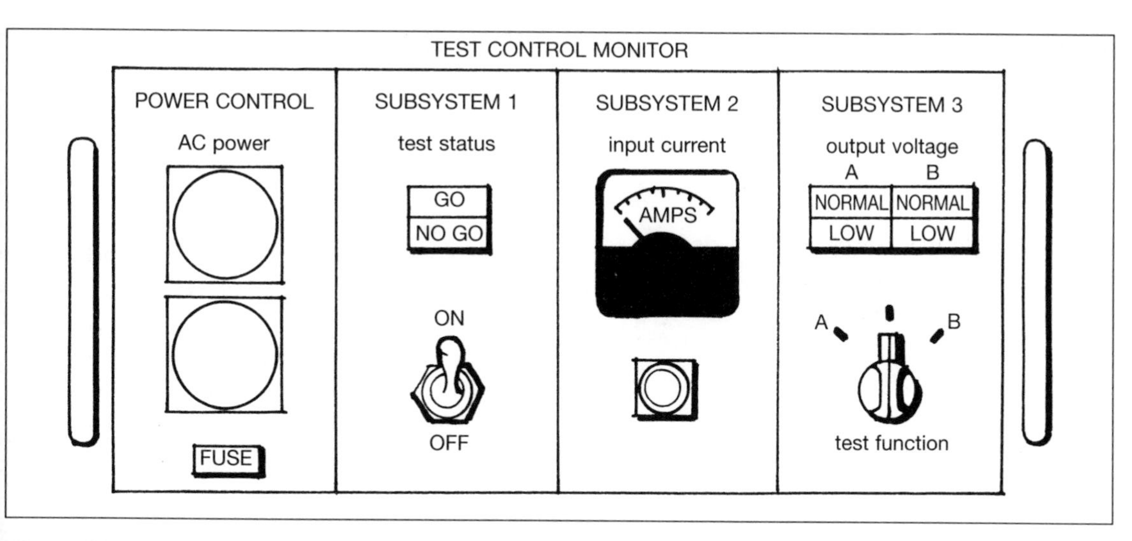

Figure 48a. Labels and signs should be easy to see and easy to read

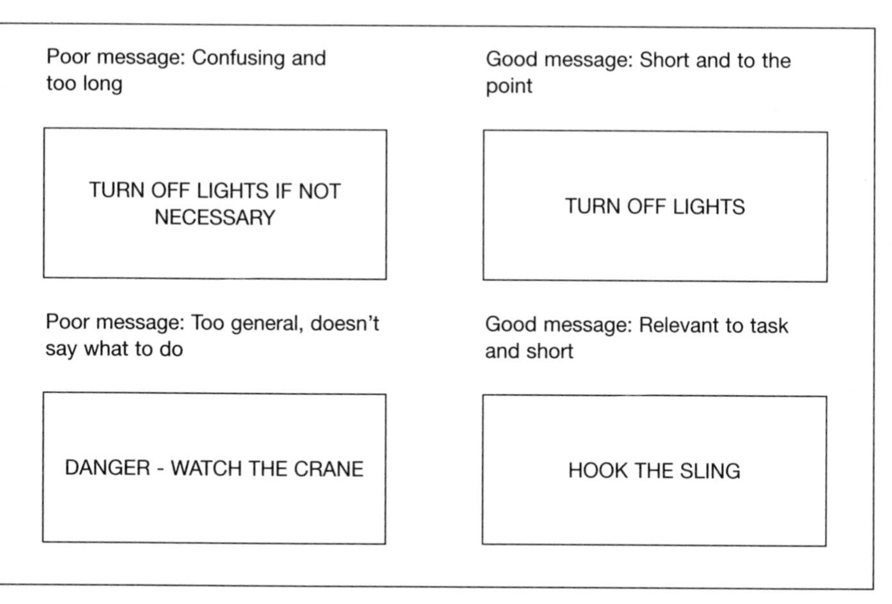

Figure 48b. Labels and signs should be short and to the point

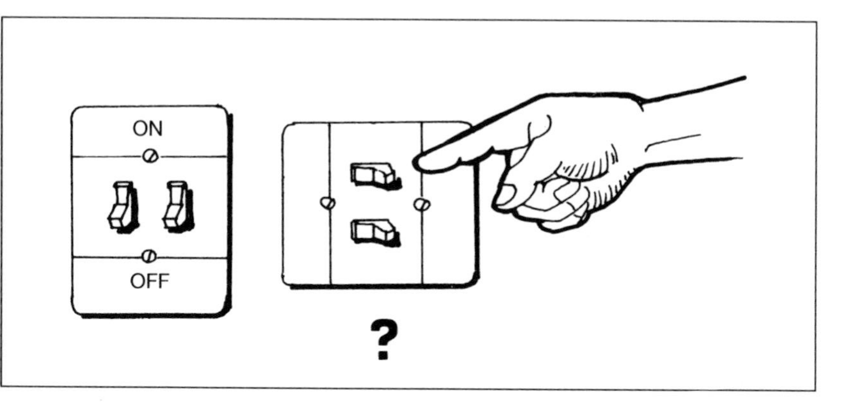

Figure 48c. Make switches easy to understand

CHECKPOINT 49

Use warning signs that workers understand easily and correctly.

WHY

Warning signs are used to warn against hazards. They often carry a complex message, since it is necessary to convey what the hazards are and what the person should do to avoid them. Make sure that warning signs are easily understood by workers.

Lengthy warning signs are in fact not read by all workers. It is important to choose compact but easy to understand messages.

HOW

1. Use a warning sign that contains four essential elements:

   (a) *a signal word* — to convey the gravity of the risk, for example, "Danger," "Warning" or "Caution." "Danger" is the most severe signal word and "Caution" the least;

   (b) *the hazard* — the nature of the hazard;

   (c) *the consequence* — what could possibly happen; and

   (d) *an instruction* — what is the appropriate behaviour to avoid the hazard.

2. Ensure that the appropriate signal word — such as "Danger", "Warning" or "Caution" — is used. Also make sure that the descriptions of the nature of the hazard and the consequences are appropriate. Check if the instruction to workers about what to do is clear enough.

3. Get workers to evaluate existing warning signs. You will gain many useful suggestions.

4. Example of a good warning sign:

   DANGER!

   HIGH-VOLTAGE WIRE

   CAN KILL!

   STAY AWAY!

SOME MORE HINTS

- Note that short messages are more effective than long ones.

- General warning signs, such as those which merely say "Danger," "Look out" or "Warning", are not effective. They are too general, and people do not understand what to do.

- Written warning signs assume that the workers are able to read. When easy-to-understand symbols are available, use both symbols and written signs.

POINTS TO REMEMBER

Warning signs must spell out what the danger is and what to do.

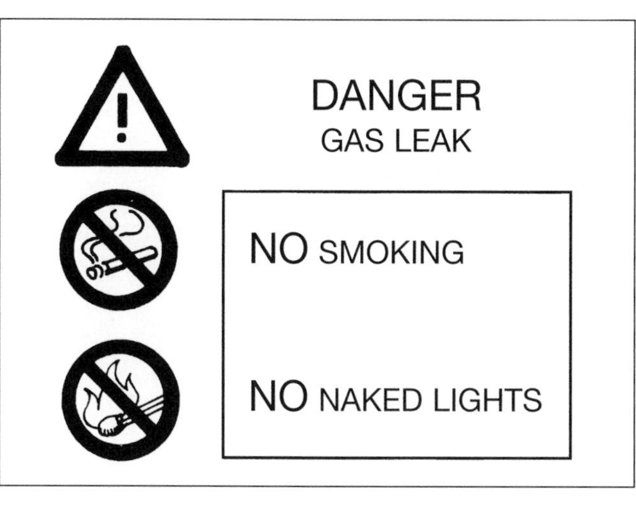

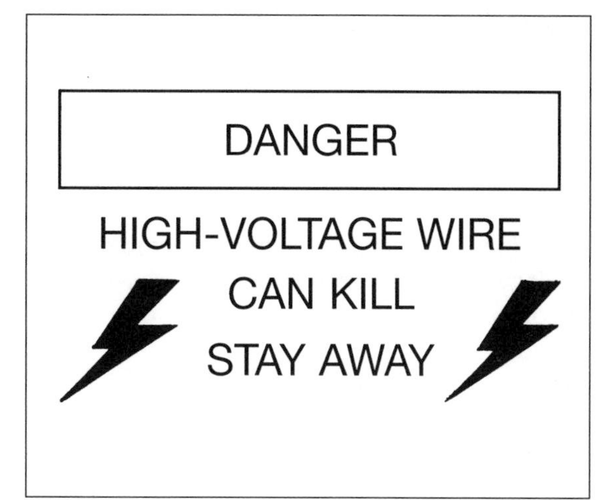

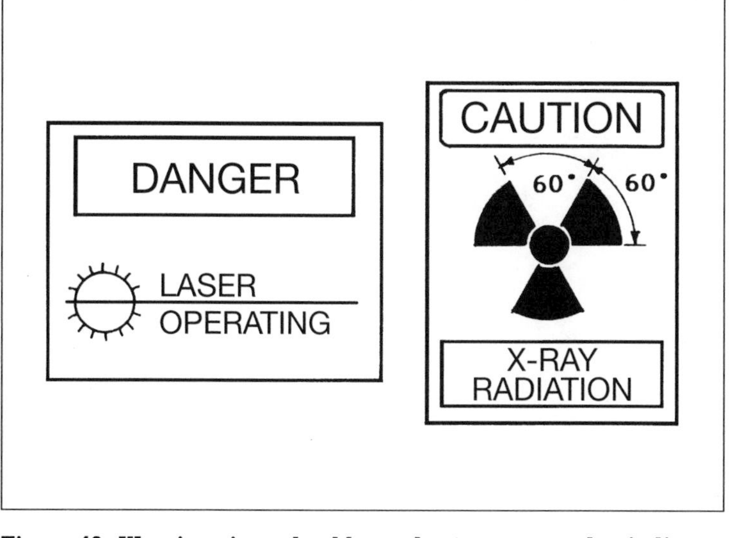

Figure 49. Warning signs should use short messages that indicate the nature and gravity of the hazards, and what to do or what not to do

CHECKPOINT 50

Use jigs and fixtures to make machine operation stable, safe and efficient.

WHY

Jigs and fixtures hold work items firmly in correct positions. They make the operation more stable and more efficient.

Fixtures leave both hands free to work.

Jigs or fixtures keep the hands away from tools or operation portions of the machine. This is because the jigs or fixtures, and not the hands, hold the work items. This increases safety and efficiency.

HOW

1. Design a jig that guides a tool or an operating part of the machine to a precise location on the work item. This will increase efficiency.

2. Alternatively, use a fixture that holds one or more items for processing. This frees the hands.

3. Always use jigs and fixtures in such a way that they firmly hold the workpiece while preventing its movement in either direction along the X, Y and Z axes, and rotation in either direction about the X, Y and Z axes.

4. Make the jigs and fixtures easy to load and unload.

5. Standardize components of jigs and fixtures (bases, bushes, pins, clamps) to minimize costs and speed repairs.

6. Establish a plan to maintain jigs and fixtures properly and make it clear to all concerned workers, so that they know what to do if the parts in the jigs or fixtures are defective (whom to contact, etc.).

SOME MORE HINTS

- As the jig or fixture weight increases, consider mechanical handling instead of manual handling.

- Chamfer sharp edges.

- Make jigs and fixtures sturdy as they tend to get rough treatment. Use wear strips on the base where they are in contact with a conveyor. Use plastic or rubber "bumpers".

POINTS TO REMEMBER

Don't use the hand as a fixture. For that purpose, use a jig or a fixture.

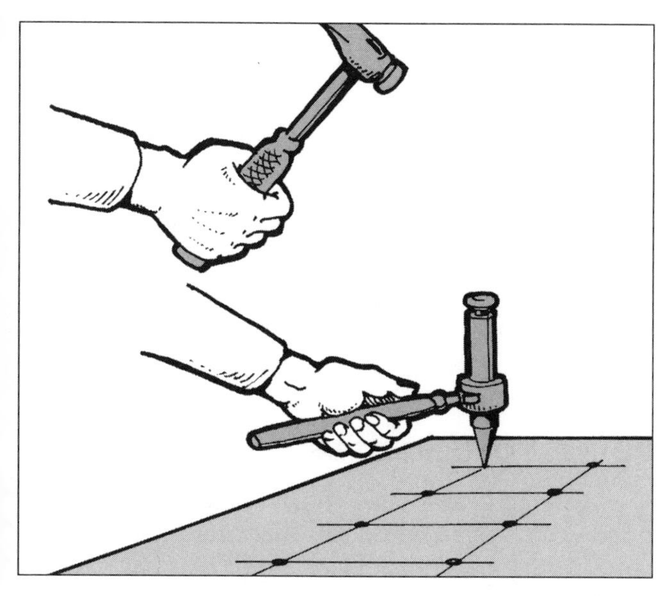

Figure 50a. Hand-held tools can be stabilized, making the job easier to perform

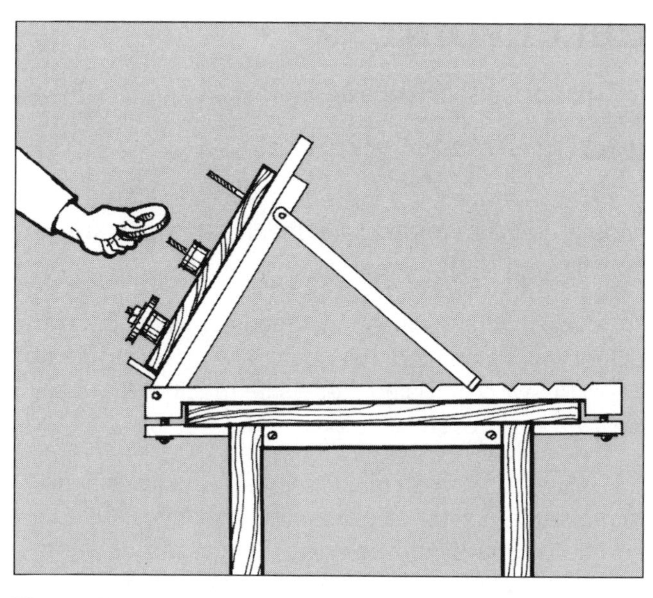

Figure 50b. Fixtures which stabilize operations can often be designed simply

Figure 50c. Machine operations can be made safer and more efficient when stabilized with a jig or a fixture

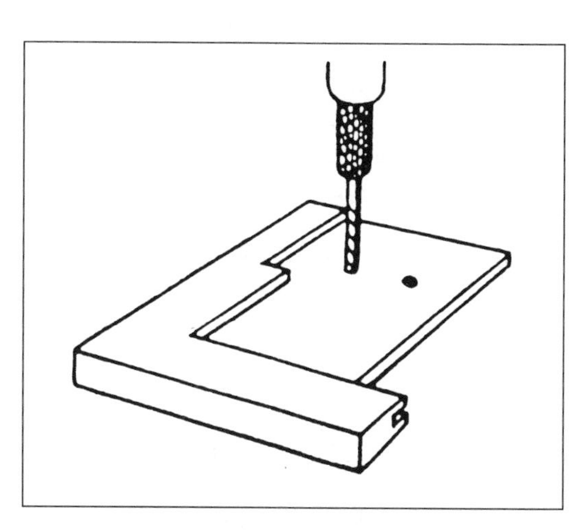

Figure 50d. Use a specially designed or universal jig or fixture instead of holding an unstable workpiece by hand

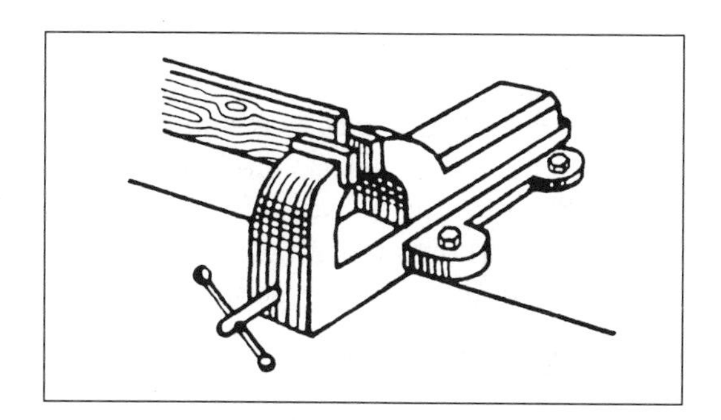

Figure 50e. Clamps and vices can hold different sizes and shapes of workpiece steady during work, and can free the hands as well

CHECKPOINT 51

Purchase safe machines.

WHY

There are safe and unsafe machines. Care should be taken to purchase machines which are constructed safely.

Safe machines mean machines in which dangerous parts are situated in a position where they cannot harm the worker. Using these safe machines is the best way to prevent accidents.

After purchasing machines, it is usually difficult to make them safer as production continues. Often additional guards or enclosing the dangerous parts can help, but it is better to purchase machines in which all these necessary guards are already in place.

HOW

1. When purchasing a machine, study the options carefully and order one in which all moving parts are guarded and points of manual operation are free from danger.

2. Confirm whether rotating shafts, wheels, rollers, pulleys and gears, as well as reciprocating motions, are adequately guarded.

3. Check if feeding and ejection can be done safely without the hands coming into a dangerous point while the machine is in motion.

4. Also check if maintenance of the machine can be done safely. In particular, the motion of the machines should be locked while they are repaired or while the maintenance work is performed.

5. Make the manual for proper operation of the machine available to all the workers concerned and provide training. Make sure that operating instructions and labels are in the language easily understood by the workers. Keep in mind that some workers may not read well or at all; providing training is essential.

SOME MORE HINTS

- Automatic or mechanical feeding and ejection devices can eliminate risks while greatly increasing productivity.

- Interlock guards are preferable because electrical or mechanical cycling of the machine is automatically interrupted if the guard or cover is opened or removed for operation or maintenance.

- You may be offered a machine without guards or unsafe versions at a lower price. Such machines can cause you many problems and cost you more in the long run. Save yourself a lot of trouble and expense by choosing the right machines.

POINTS TO REMEMBER

Working in fear of accidents greatly hampers good work results. Install safe machines that cannot harm workers. Safe machines are productive machines.

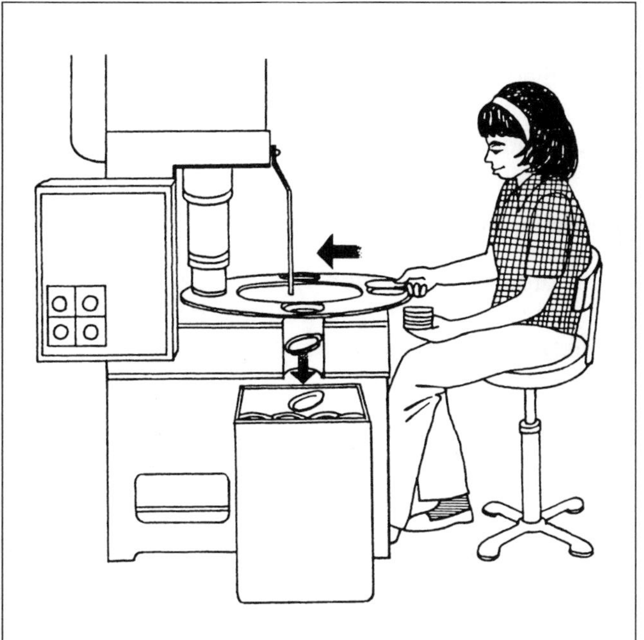

Figure 51a. Power press with carousel feed

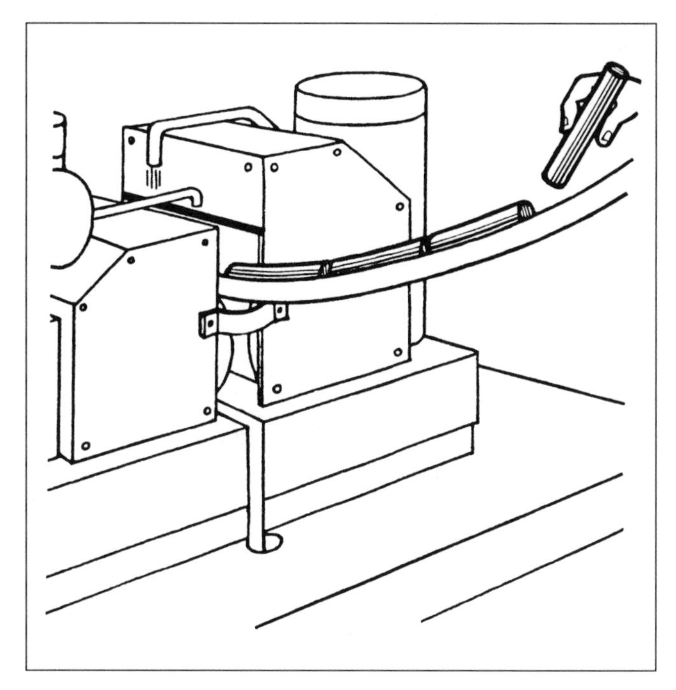

Figure 51b. A machine with a self-feeding device keeps the workers' hands away from dangerous parts of the machinery

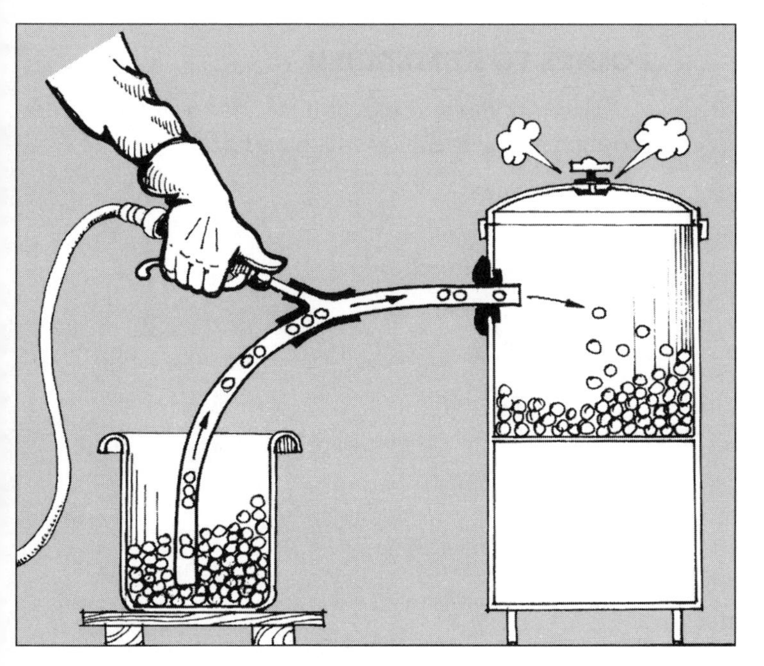

Figure 51c. Semi-solid or granular materials can be moved with pressurized air

CHECKPOINT 52

Use feeding and ejection devices to keep the hands away from dangerous parts of machinery.

WHY

With feeding and ejection devices, objects can be handled with greater precision and without risk of injury.

Feeding and ejection devices can greatly reduce the time for feeding and unloading. Using the time saved, the worker can carry out other tasks, such as preparing for the next work item. This means less idle time for the machine.

The use of feeding and ejection devices makes it possible to remove wastes or toxic substances without handling them manually.

HOW

1. There are many different types of feeding/ejection device. The following are some examples of simple types:

   - *plunger feed:* a plunger with a die (a slot or nest) into which the stock is placed outside the point of operation and then pushed into the point of operation as the machine is cycled;

   - *carousel feed:* a carousel type of feeder is one in which the stock is placed outside the point of operation and put under the point of operation one item at a time, combined with automatic ejection and collection of finished stock;

   - *gravity chute feed:* automatic placing of the stock in the point of operation or in the plunger device, thus saving the worker from having to place new stock at each cycle.

2. Use compressed air for feeding semi-solid or granular materials.

3. Use an ejection device that is part of the feeding system. This saves ejection time. When a separate ejection device is needed, use a mechanical device or compressed air.

4. Use feeding aids, such as hooks, bars or other extensions, to feed or remove objects. In each individual case, an appropriate solution must be invented. For example, use a hook with a rounded handle to remove cutter shavings from a turning lathe.

SOME MORE HINTS

- There are many other ways to benefit from "free" gravity. In some cases, a simple inclined chute feeder can be used to move the stock into the point of operation.

- The feeding and ejection devices must not interfere with existing guards or other safety devices.

- The maintenance of feeding and ejection devices or removal of an operation failure must not cause an inadvertent cycling of the machine.

- The correct height and the placement of the feeding devices make work easier and more efficient.

POINTS TO REMEMBER

Use feeding and ejection devices to increase productivity and reduce machine hazards.

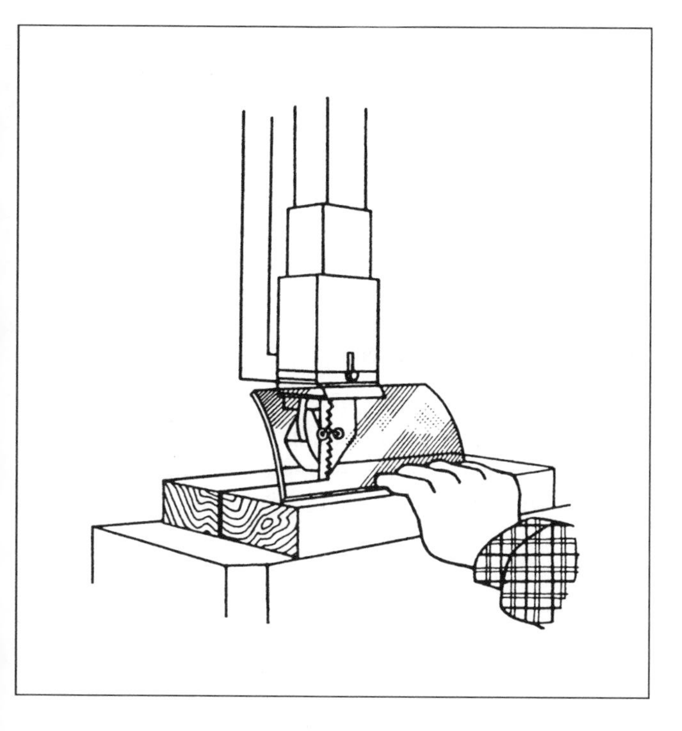

Figure 52a. An adjustable guard for a bandsaw

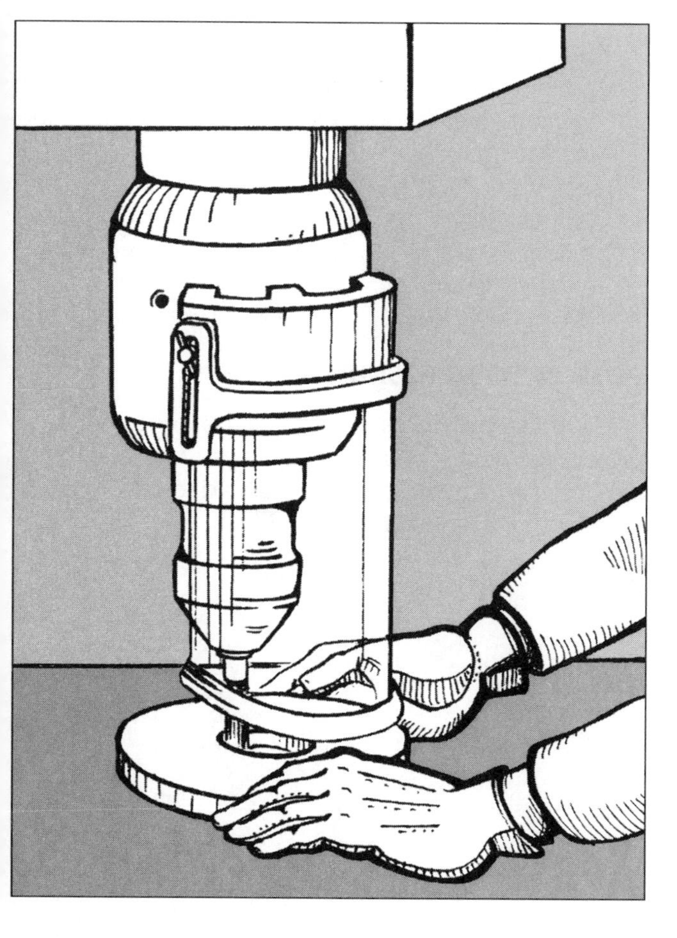

Figure 52b. A well-designed machine guard should prevent contact with moving parts of the machine and should allow the worker to do his or her job comfortably

CHECKPOINT 53

Use properly fixed guards or barriers to prevent contact with moving parts of machines.

WHY

When working near moving parts of a machine, workers are at risk. Injuries may occur from the power transmission parts (such as gears, shafts, wheels, pulleys, rollers, belts or hydraulic lines), from the point of operation or from flying objects such as chips, sparks or hot metal. The best protection against the risk is by preventing contact through mechanical means, not by instructing workers to avoid it.

Accidents may happen during normal operation or during cleaning and maintenance. Often bystanders and other workers can be at risk, since they do not understand how the machine operates and what precautions are necessary. Observe the national standards that prescribe the use of machine guards and barriers, and improve on these further to protect people.

HOW

1. Design a fixed guard that can be attached to the machine for protection against both the machine itself and flying objects. The guards must be practical to use. They must meet the requirements of the machine and the specific danger.

2. If the machine guard hinders manual operation or if workers cannot see the task clearly, workers will most likely remove the guard. Redesign these guards or replace them with adjustable guards that can be adjusted to suit the size of work items being introduced into the point of operation and still provide a high degree of protection.

3. To make it possible to see the task clearly, use machine guards made of plastic or see-through material.

4. Put up fixed barriers in places where contact with moving parts of the machine is possible, even though this danger is not readily visible. Make sure that these barriers are stable and high enough for the purpose.

5. Where one moving part comes into contact with another and thus makes up a "pinch point", put up fixed barriers or appropriate guards to prevent fingers or hands from being caught.

6. Similarly, when two rotating rollers roll together and thus make up a "nip point", erect appropriate guards to prevent hands or clothing from being caught.

SOME MORE HINTS

- Guards may be attached directly to the machine or to a stable surface such as a wall or a floor. They should be made of strong material and provide protection against flying fragments.

- Fixed guards should be removable only by using tools.

- Fixed guards at the point of operation should be accompanied by appropriate feeding and ejection devices so as to facilitate safe operation and increase efficiency. Special hand tools may also be used to reach into the point of operation and manipulate work items (e.g. pliers and tongs with vacuum suction devices or magnetic lifters at the end).

- Manufacturers of machines usually supply machine guards. Sometimes these are impractical and you may find that it is necessary to design your own guards.

POINTS TO REMEMBER

Machine guards and barriers are important to protect workers and bystanders. If you find that they are not used, immediately seek an adequate solution by erecting a redesigned guard.

Figure 53a. A machine with two-hand controls

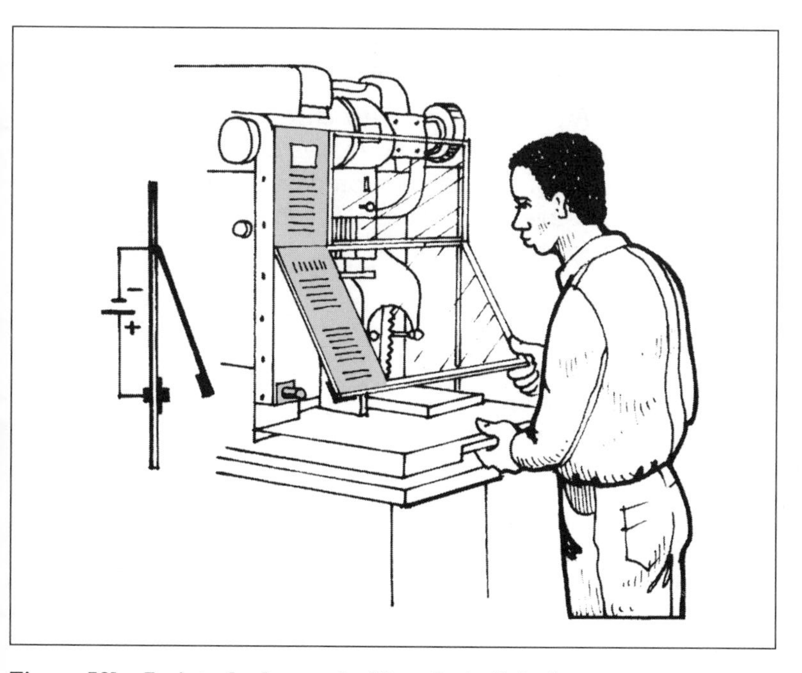

Figure 53b. An interlock guard with a shut-off device

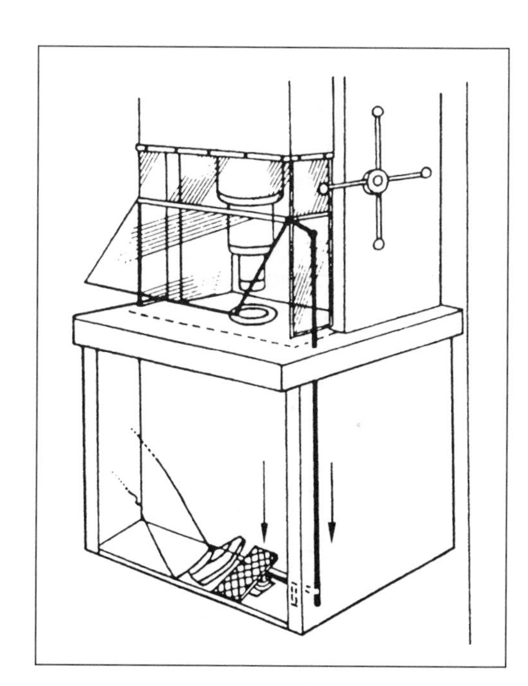

Figure 53c. Pedal activation of an interlock guard

CHECKPOINT 54

Use interlock barriers to make it impossible for workers to reach dangerous points when the machine is in operation.

WHY

Accidents happen quite often when the worker opens or removes the guard or cover. If the machine stops when the guard or cover is opened or removed, there is no danger.

Interlock guards or barriers automatically interrupt the electrical or mechanical cycling of the machine as soon as the guard or cover is opened or removed.

Interlock systems may also block access to the point of operation just prior to the work cycle.

HOW

1. Construct a fence with a gate to enclose the process. An interlock barrier typically requires a key to open the gate. When the gate opens, an automatic switch turns off the power supply to the machine. The interlock gate needs to be closed before the dangerous machine starts moving again.

2. Where mechanical interlocking is difficult to apply, use photosensitive interlock systems. They interrupt the machine operation whenever any part of the body is beyond "light barriers" that have light sources on one side and light-receiving parts on the other.

3. Great care must be taken when a process continues in its cycle, to see that it takes more time to open the gate than the process takes to stop.

4. If interlocking is not possible, two-hand controls can be used. Two-hand controls require that two switches or levers must both be operated at the same time with both hands. In this way, the worker's hands are always outside the machine while it operates.

SOME MORE HINTS

- Because interlocks or two-hand controls may be inconvenient for the production process, they are sometimes tampered with. The interlocks and their switches should be designed so that they are tamperproof and cannot be easily broken or overridden with screwdrivers, pencils or adhesive tape. Two-hand controls should be designed so that the two switches cannot be operated with one hand, taped or jammed on, pressed with the knee or otherwise circumvented.

- A large space behind the interlocking barrier can cause a serious hazard, because it is possible to close the gate behind a worker inside the danger area. Somebody else, being unaware of the presence of the worker inside, may close the gate and thus activate the machine. The key should therefore be used both for closing and opening and the worker should be told to put the key in his or her pocket so that no one else can use it while inside the danger area,

- Interlocks are also common on electrical equipment. The process equipment may be enclosed in a box with an opening and a key. The key opens up the box and breaks the power supply.

POINTS TO REMEMBER

An interlock is an effective means to protect workers from the danger area of a machine. It is used to turn off a production process automatically, making it possible for workers to reach work items, inspect or repair.

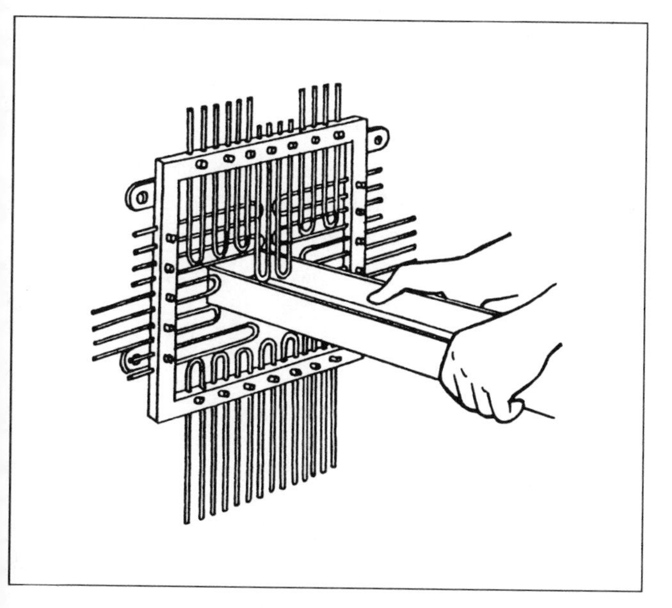

Figure 54a. An adjustable guard on a power press

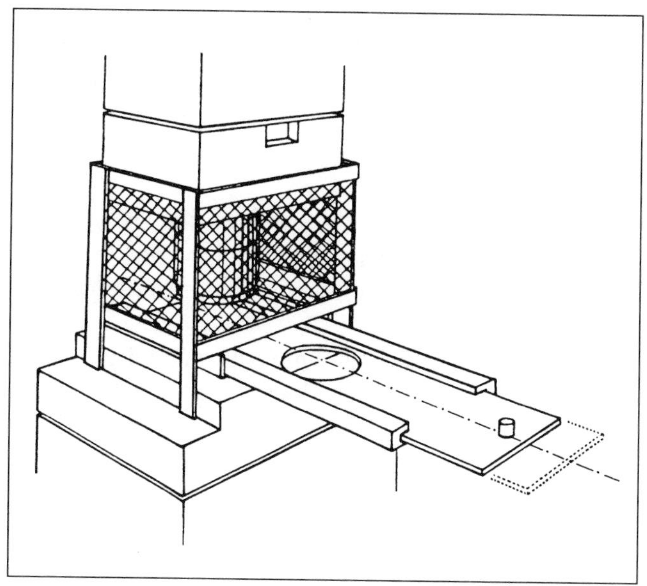

Figure 54b. A power press with plunger feed

CHECKPOINT 55

Inspect, clean and maintain machines regularly, including electric wiring.

WHY

A well-maintained machine is less likely to break down. A poorly maintained machine can not only have more breakdowns but can also be dangerous.

A well-maintained machine with safe wiring is less likely to catch fire and electrocute workers.

Machine guards also should be inspected, cleaned, and repaired or replaced, as necessary.

HOW

1. Develop a schedule of routine inspection, cleaning and preventive maintenance.

2. Create an inspection and maintenance log (record book) for each machine and each work area. Make this log available to all workers.

3. Designate key personnel to be responsible for inspecting the machines and the logs.

4. Maintenance should also include seeing that all necessary machine guards are in place.

5. Train workers to perform inspections at their own work area and report deficiencies.

6. When machines are being repaired or when maintenance tasks are being performed, the control mechanisms of the machines should be locked and should have a tag saying "DANGER! DO NOT OPERATE!".

SOME MORE HINTS

- A machine maintenance programme, carried out by qualified personnel, will reduce the frequency of repairs and reduce the need for the worker to remove guards.

- Cooperation of all workers is necessary for proper maintenance and cleaning of machines. Make it clear that the maintenance programme is an essential part of good production management.

- Reward workers for inspecting and maintaining the machines.

POINTS TO REMEMBER

Proper maintenance does not mean lost production time. It is an investment in higher productivity, lower repair costs and safety.

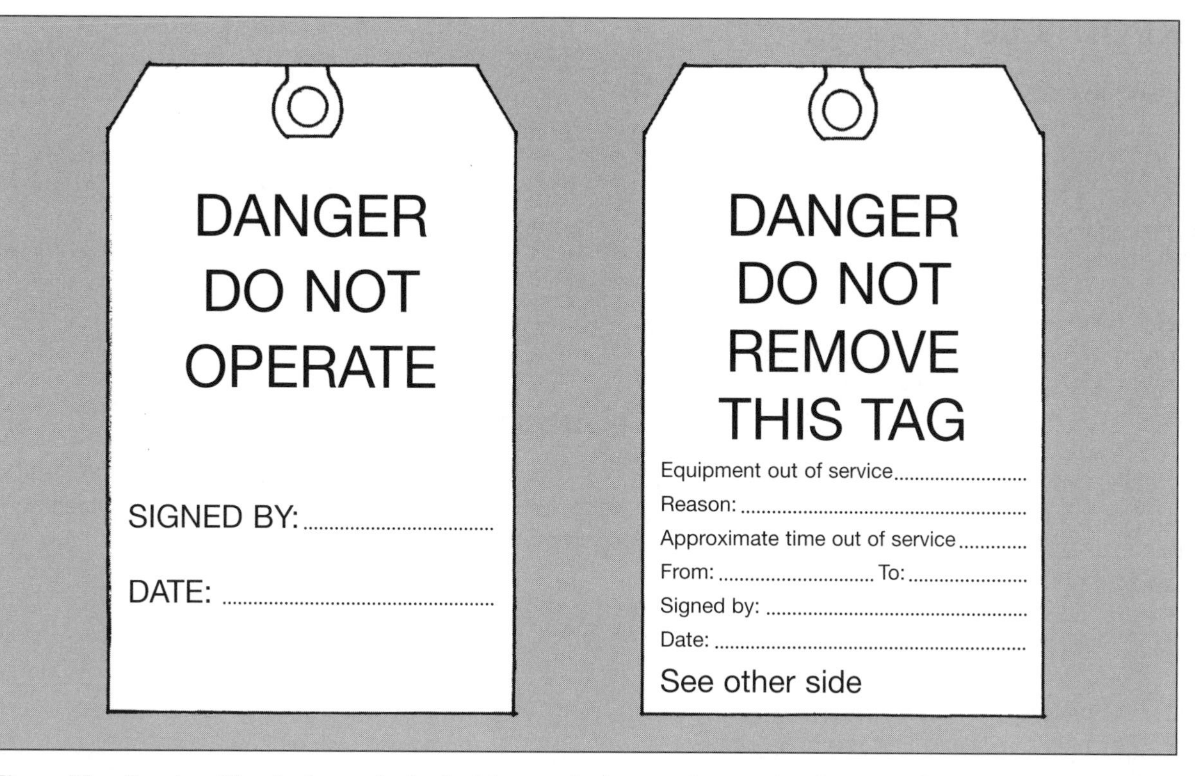

Figure 55a. Front and back views of a typical tag used when equipment is taken out of service because it has become unsafe

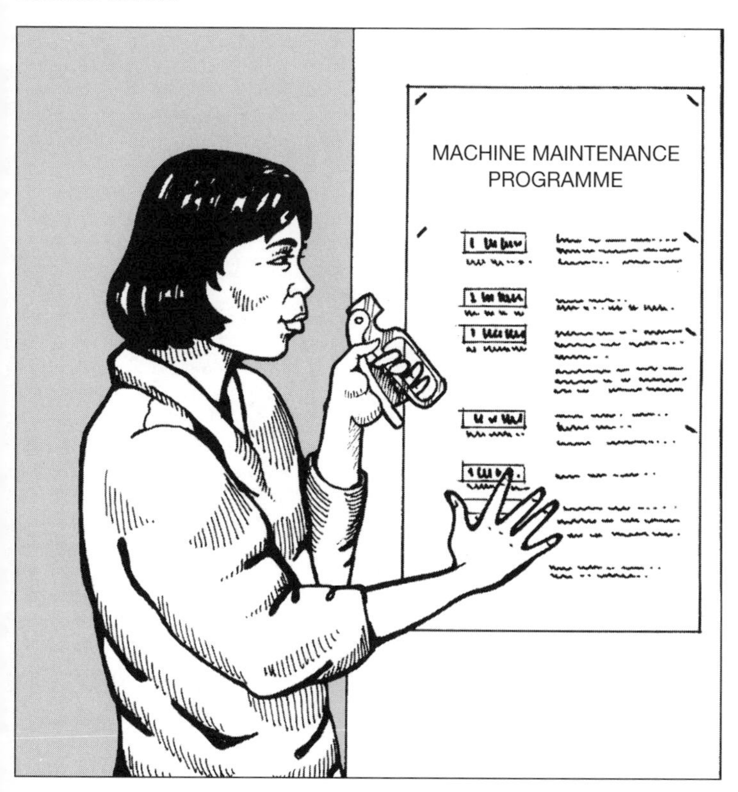

Figure 55b. Make known to all workers concerned the machine maintenance programme in operation and train them to perform their duties

CHECKPOINT 56

Train workers for safe and efficient operation.

WHY

Training and retraining of workers in safe and efficient operation are an indispensable part of daily production.

Machines have advantages over people in strength, speed and accuracy. To realize these advantages fully, workers must be trained to use machines safely and efficiently.

Improper use of machines can cause work slowdowns, work stoppages, damage and injury. These events entail costs which must be added to the already high cost of the machinery that was not used properly.

HOW

1. Establish training programmes involving all workers. For newly recruited workers, organize training sessions that include training in safety and efficient production.

2. In training workers, use information provided by the manufacturer of the machines and equipment. Translate this information into local languages. If it is too complex, use step-by-step procedures.

3. Involve workers who are already proficient in using the machines in the training, in particular by letting them demonstrate proper and safe operation. When workers master skills, let them coach new workers.

4. Use good examples (pictures, videos, demonstrations) as models for others to learn from.

SOME MORE HINTS

- You can obtain detailed information about the machinery from the manufacturer or the company that sold you the equipment. To know about these companies, seek help from trade associations, your dealers, fellow companies or available registry books.

- Train workers on new machines when they are brought into the enterprise, not after problems appear.

- You may want to trace errors, accidents or defects to particular machine operations. These may be eliminated with proper training using the local language and relevant examples.

- If you make up your own training materials, remember to:
  - use simple ideas;
  - use pictures or drawings to illustrate points;
  - make sure that the words and language are clear to workers.

POINTS TO REMEMBER

Training workers how to use machines is paid for once. Errors, rejections and scrap losses, injuries and low quality are paid for forever.

Figure 56a. Train workers to use machines properly and safely

Figure 56b. (i) and (ii) In training workers, include good visual examples and learning by doing

Improving workstation design

CHECKPOINT 57

Adjust the working height for each worker at elbow level or slightly below it.

WHY

The correct height of places where work is done with the hands facilitates efficient work and reduces fatigue. Most work operations are best performed around elbow level.

If the work-surface height is too high, the neck and shoulders become stiff and painful as arms must be held high. This happens in both standing and sitting positions.

If the work surface is too low, low back pain easily develops as the work has to be done with forward bending of the body. This is serious in a standing position. In a sitting position, a too low working height causes both shoulder and back discomfort in the long run.

HOW

1. For seated workers, work surface height should be around elbow level. When some forces are exerted downward, then the working height should be slightly below elbow level. If a keyboard is used, the keyboard height at which the fingers operate should be at or slightly below elbow level.

2. An exception should be made for high-precision work while sitting. In this case, the object can be raised slightly above elbow level to allow the worker to see the fine detail. In this case, provide armrests.

3. For standing workers, the hand height should be a little or somewhat below elbow level. For work requiring accuracy, elbow height can be chosen. In light assembly work or packing of large items, the hand height should be about 10-15 cm lower than elbow level. When the use of very strong force is needed, an even lower height is appropriate so as to allow the use of body weight. However, too low a work height that causes low back pain should be avoided.

4. Where possible, use an adjustable work table, for example a lift-table with a hydraulic device for raising or lowering the table height.

5. Use a wooden platform or a similar flat structure under tables, work surfaces or work items to raise the working hand height. Use platforms under the feet or under the chair to lower the actual working height in relation to elbow level. These adjustments are extremely effective.

SOME MORE HINTS

- Adjusting working height is much easier than people normally think. As machines or tables are involved, people tend to think that changing work height is impossible or too expensive. This is not true. Learning from the above examples, use your own ideas.

- Adjustable work tables are available. They facilitate the use of the same workstation by several people, and thus increase productivity.

- If the same work table is used for both standing and sitting work, particular care is needed to provide a higher working surface for standing work and to avoid too high a working height for seated work. This is usually solved by choosing a table suited to seated workers and by inserting platforms or fixtures under work items handled while standing to give higher levels for standing workers. Alternatively, choose a table height for standing work and provide high chairs and adjustable footrests for seated work.

POINTS TO REMEMBER

Apply the "elbow rule" to determine the correct hand height for greater efficiency and for reducing neck, shoulder and arm discomfort.

Figure 57a. Most work operations are best performed around elbow level

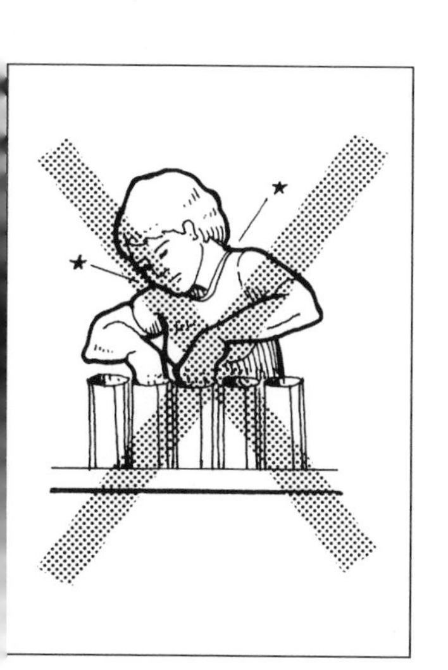

Figure 57b. For seated workers, work-surface height should be around elbow level

Figure 57c. Recommended dimensions for most seated tasks

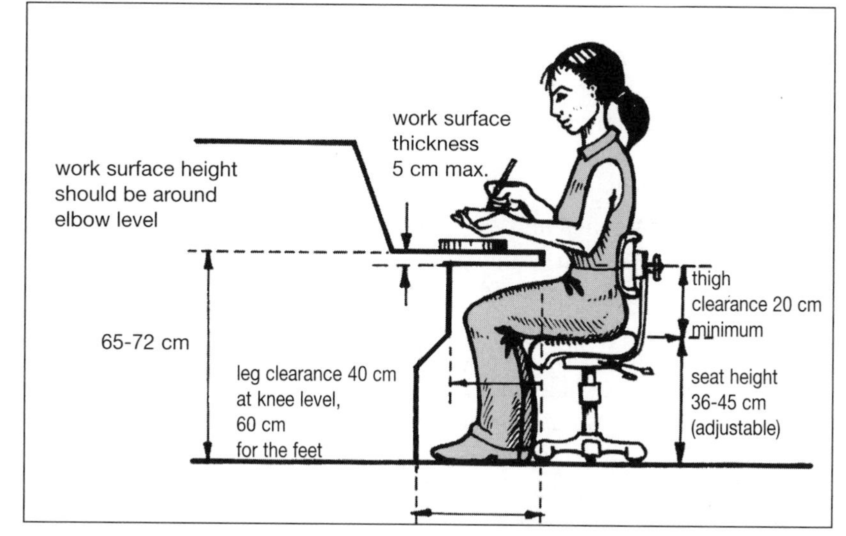

Figure 57d. Recommended dimensions for standing work

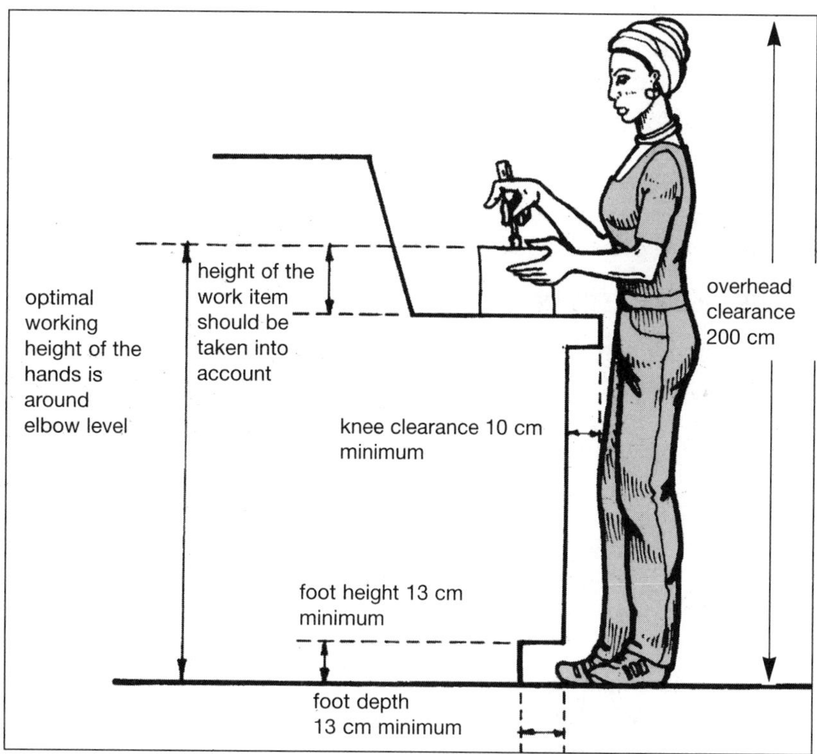

CHECKPOINT 58

Make sure that smaller workers can reach controls and materials in a natural posture.

WHY

Differences in body size of workers are usually very large in any workplace. The differences are becoming even larger with time as workers of both sexes from different regions come together. Special care must be taken so that controls and materials can be reached easily by all workers.

Controls and materials that are too remotely placed fatigue smaller workers and reduce their efficiency. This is dangerous and must be avoided.

HOW

1. Purchase machines and equipment with adjustable work-surface height. Then adjust the height to suit the smaller workers.

2. Replace controls (though this might be relatively difficult once machines are bought) and materials so that they are within easy reach of smaller workers. If the same controls and materials are dealt with by taller workers, make sure that they are still within easy reach of the taller workers.

3. Use platforms for smaller workers so that the hand position of these workers becomes higher and can easily reach controls and materials.

4. Use a foot-stand or a mobile platform to enable workers to reach particular controls or materials which are difficult for them to reach.

SOME MORE HINTS

- Ask smaller workers if they have difficulties in reaching controls and materials. Discuss with them how this can be improved. There are usually practical ways to solve the problem.

- For a lever-control, an extension can make it easy for smaller workers to operate it. Consider similar arrangements to improve the difficult reach of some controls.

- A mobile control panel or keyboard can make the workstation easily adjustable for both larger and smaller workers.

POINTS TO REMEMBER

Make sure that smaller workers can reach controls and materials naturally.

Figure 58a. Use foot platforms for smaller workers to ensure an appropriate work height at around elbow level

Figure 58b. Difficult reaching should be avoided

CHECKPOINT 59

Make sure that the largest worker has enough space for moving the legs and body easily.

WHY

Generally, adjustment of work-surface height for larger people is relatively easy. However, clearance for movement or clearance under the work table is difficult to expand once the workstation is installed. Clearance must be large enough from the outset to accommodate larger people.

In order to accommodate larger people, it is most important to provide adequate leg and knee clearance. Extra space is also necessary to accommodate taller people.

Enough space to move the legs and body easily will reduce fatigue and the risk of musculoskeletal disorders, thus improving the worker's efficiency.

HOW

1. Check overall space clearance of all workstations and passageways for the largest worker, and increase clearance where necessary.

2. Check knee and leg clearance of workstations used by the largest worker. If knee and leg clearance is too narrow, consider how the clearance can be expanded. Raise the work-table height or expand the work-table size, for example.

3. Mark all unsafe clearances with bright colours and warning signs.

SOME MORE HINTS

- Ask the largest worker where he or she feels unsafe or whether the space is too narrow. Take measures to deal first with unsafe conditions, and then with discomfort.

- It is uneconomical and impractical to design equipment for people of all sizes. Often equipment is designed to accommodate about 90 per cent of the proposed user population, which means that the smallest and largest 5 per cent may be excluded. In your workplace, therefore, make sure that even the largest and smallest workers feel safe and comfortable with the existing space. Just following regulations might not be enough.

- Also consider the other body-size related needs faced by larger workers: gloves, protective clothing, helmets, etc.

POINTS TO REMEMBER

Make sure that the largest workers feel comfortable and safe with the existing space.

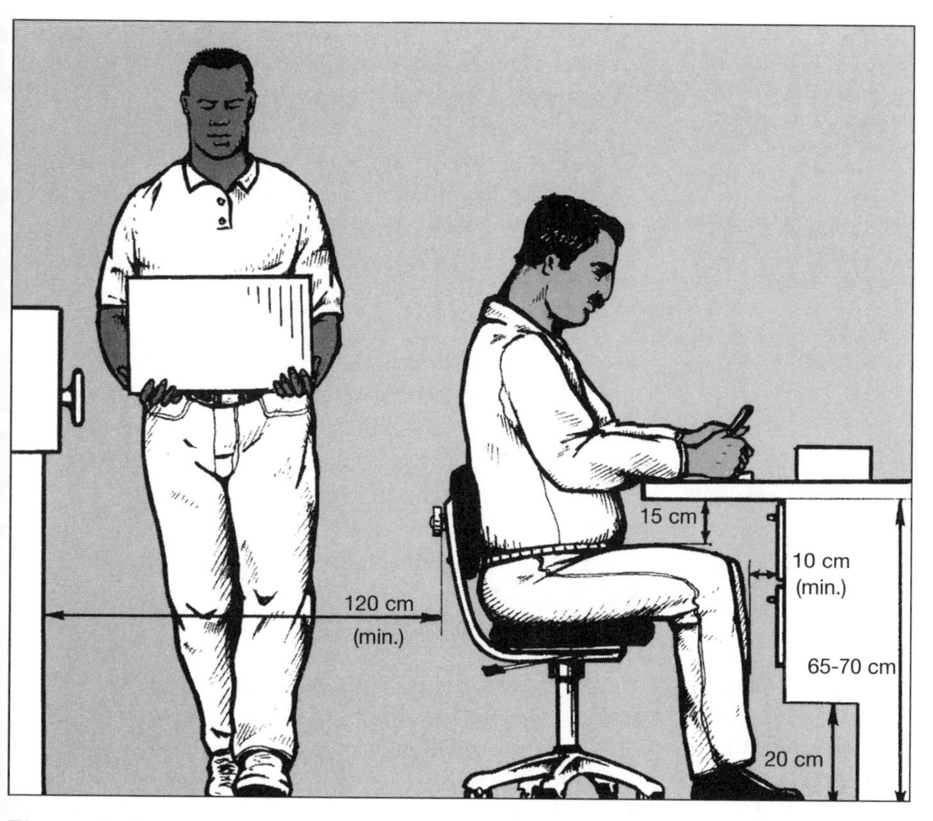

Figure 59. Ensure enough space for larger workers in aisles and at workstations. Knee and leg clearance should not be forgotten

CHECKPOINT 60

Place frequently used materials, tools and controls within easy reach.

WHY

Time and energy are saved by placing materials, tools and controls within easy reach of the workers.

Long reaches mean a loss of production time and extra effort.

The distance which can be reached easily without bending forward or stretching is quite small. Long reaches can thus lead to neck, shoulder and back pain, as well as to imprecise operations.

HOW

1. Place frequently used tools and controls within the primary hand movement area (between 15 and 40 cm from the front of the body and within 40 cm from the side of the body at elbow height).

2. Place all the frequently used materials within this primary hand movement area or at the margin of this area. When materials are supplied in boxes or bins, or on pallets or racks, they should be placed within easy reach and at around elbow height.

3. For similar workstations, organize the placing of tools, controls, materials and other work items in a good combination with each other. For example, when several kinds of material are collected at the same time or one after another, place them in the same area in different bins. Standardize the location of all these items based on the opinions of the workers.

4. If appropriate, divide the work-table surface into subtask areas so that different operations are done sequentially.

SOME MORE HINTS

- It is very important to place within this primary hand movement area all the items used regularly. Let workers adjust the workstation to their needs.

- Displays and instructions can be placed beyond this easy-reach area as long as they are presented in an easy-to-read form.

- Materials, tools and controls can be placed not only on the main work table but also on a side-table or a rack placed within easy reach. Less frequently used items can be placed at the side of the worker.

- Tools or materials used only occasionally (a few times per hour, for example) may be placed at a distance reached by leaning forward or stretching aside, or even outside the immediate work area without much loss in productivity.

- Provide adjustment for left-handed workers.

POINTS TO REMEMBER

Place frequently used materials, tools and controls within easy reach. This easy-reach area is quite narrow and you can determine it by trying to reach while keeping your natural posture.

Figure 60. Place frequently used materials, tools and controls within easy reach

CHECKPOINT 61

Provide a stable multi-purpose work surface at each workstation.

WHY

Work at any workstation consists of a variety of tasks including preparation, principal operations, recording, communication and maintenance. A stable work surface of a certain size is needed to accommodate not only principal tasks but also various other tasks.

A work surface that is too narrow or unsteady results in time loss and more effort, thus reducing work efficiency and increasing fatigue.

HOW

1. At each workstation provide a stable work surface of appropriate size where a variety of tasks can be done, including preparation, main tasks, recording, communication and maintenance-related tasks. Such a surface is usually available when the work requires a work table, but tends to be neglected when the main operations do not require a table.

2. Avoid a makeshift work surface or an unsteady surface. Work done on it becomes frustrating and of low quality.

3. The thickness of the work surface should be not more than 5 cm. This is necessary to secure knee space underneath. Therefore avoid putting drawers or under-table shelves in front of the seated worker where the legs are positioned.

4. In the case of a visual display unit (VDU) workstation, a work surface is needed, in addition to the keyboard space, for preparation, document holding, writing and maintenance.

SOME MORE HINTS

- Consider the whole working day at the workstation. Pay due attention to all the necessary preparatory and subsidiary tasks. A work surface of a certain size is often useful even if the main tasks do not necessitate it.

- Also consider places for small tools, stationery and other personal items.

- If appropriate, use a side-table, an existing flat surface on a rack or nearby workstands.

POINTS TO REMEMBER

Provide a stable work surface at each workstation for use for a variety of preparatory, main and subsidiary tasks.

Figure 61. Provide a stable multi-purpose work surface at each workstation

CHECKPOINT 62

Provide sitting workplaces for workers performing tasks requiring precision or detailed inspection of work items, and standing workplaces for workers performing tasks requiring body movements and greater force.

WHY

Selection of a sitting or standing workplace usually depends on tradition and experience. By carefully examining which of them is better, you have a good chance of improving productivity and quality of work. Bad work postures can cause back, shoulder, neck and arm disorders.

Sitting work is more suited to precision work, while standing work is suited to various other kinds of manual work. Since working height is usually different between sitting and standing tasks, design the workstation according to the nature of the work.

Sitting all the time or standing all the time is tiring. It is always better to provide opportunities to alternate between sitting and standing. In so doing, we should know which tasks are suited for sitting and which others for standing.

HOW

1. Check jobs in which workers complain of fatigue or discomfort. Identify the principal tasks for each such job and find out if the sitting or standing workstation used is suitable.

2. Provide a sitting workstation for a job requiring high accuracy, repetition of detailed manipulation or continued watchkeeping. Use workstation dimensions suitable for sitting work.

3. Provide a standing workstation for a job requiring a lot of body movements and greater force. Experienced workers can easily tell if such a workstation is preferable. Use workstation dimensions suitable for standing work.

4. Make sure that enough room is provided under the work surface for knees, legs and feet.

5. Make the workstation adjustable, in particular its working height. If this is not possible, choose a suitable height for each individual worker, e.g. by providing an adjustable chair, a platform, etc.

SOME MORE HINTS

- A useful basic principle for determining the appropriate working height is to operate at or a little below elbow level.

- The height of the work item should be taken into account in choosing the height of the work table.

- It is desirable to assign work tasks so that the worker can alternate sitting and standing while at work. If the principal tasks do not allow this, combination with other tasks may be considered.

POINTS TO REMEMBER

Recommended dimensions for standing and sitting work are different. Provide workstations suited to the main posture.

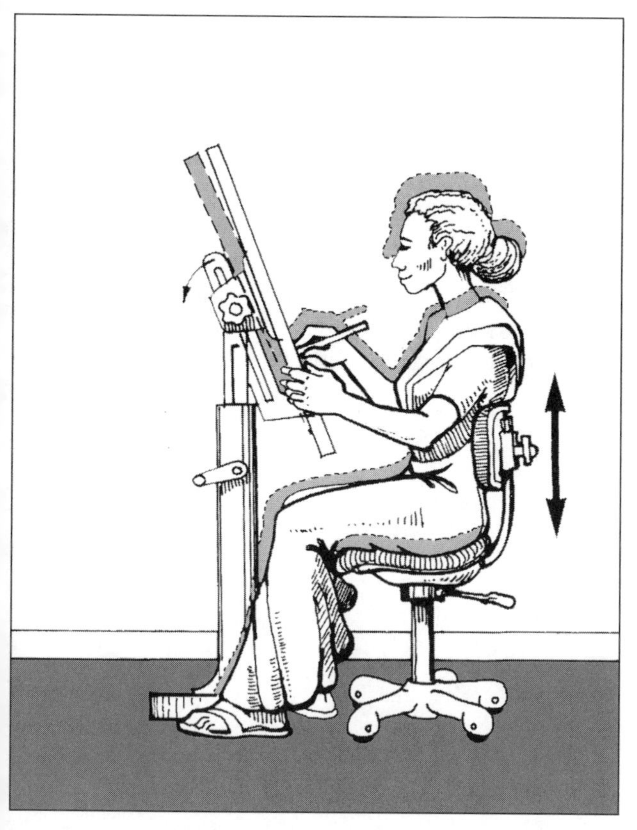

Figure 62a. Provide a sitting workstation for a job requiring high accuracy

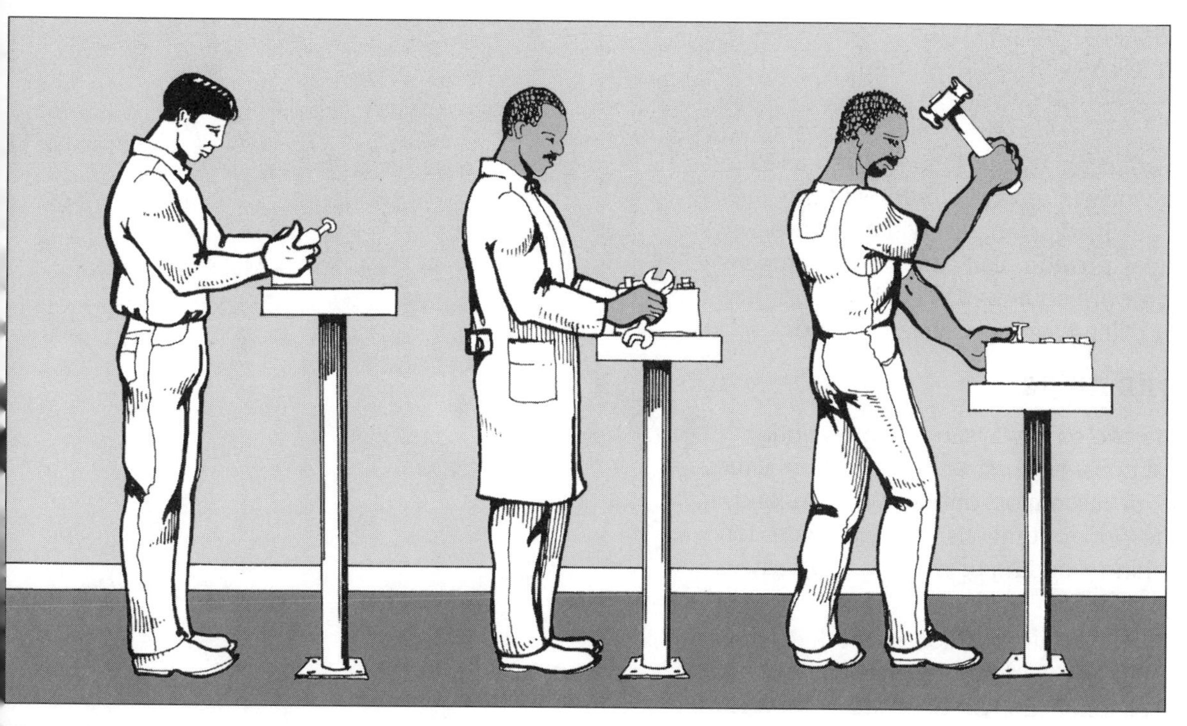

Figure 62b. Provide a standing workstation for a job requiring a lot of body movements and greater force

CHECKPOINT 63

Make sure that the workers can stand naturally, with weight on both feet, and perform work close to and in front of the body.

WHY

Working operations are more stable and efficient when done close to and in front of the body in a natural posture. The workstation should be designed to allow for such operations.

Working in an unstable position might cause a costly mistake.

Fatigue of workers, and the risk of neck, shoulder, arm and back disorders, are reduced when the work is done avoiding unnatural postures.

HOW

1. Arrange all important and frequent operations so that they are carried out close to and in front of the body, and around or slightly below elbow level. Make sure that the work table or working height close to and in front of the body is free from obstacles.

2. Make sure that these frequent operations can be performed without raising the elbow high or bending or twisting the body long enough to cause discomfort.

3. Provide adjustable workstations when used by different workers or where different tasks are carried out. If adjustable workstations are impractical, provide platforms or other means to adjust the working height to each worker. Use lifting and tilting arrangements if needed.

SOME MORE HINTS

- There are two easy ways to find out about unnatural postures. First, ask the workers if they feel pain or discomfort during work. Second, watch the work operations and find those done by stretching, bending or twisting the body.

- The optimal heights for frequent work operations are: for standing work between waist level and heart level; for sitting work between elbow level and heart level.

- Workers get tired if work operations are always done at the same place, even at the optimal place. Variations in work posture are essential. Therefore avoid repetitive tasks that need to be done in the same posture all the time.

POINTS TO REMEMBER

When work is done in a natural posture, with weight on both feet and without bending or twisting, fatigue is less and productivity is higher. Arrange for good hand positions allowing this posture.

Figure 63a. Doing work in front of the body at elbow level or at a level a little below it is always desirable. Use lifting and tilting arrangements if needed

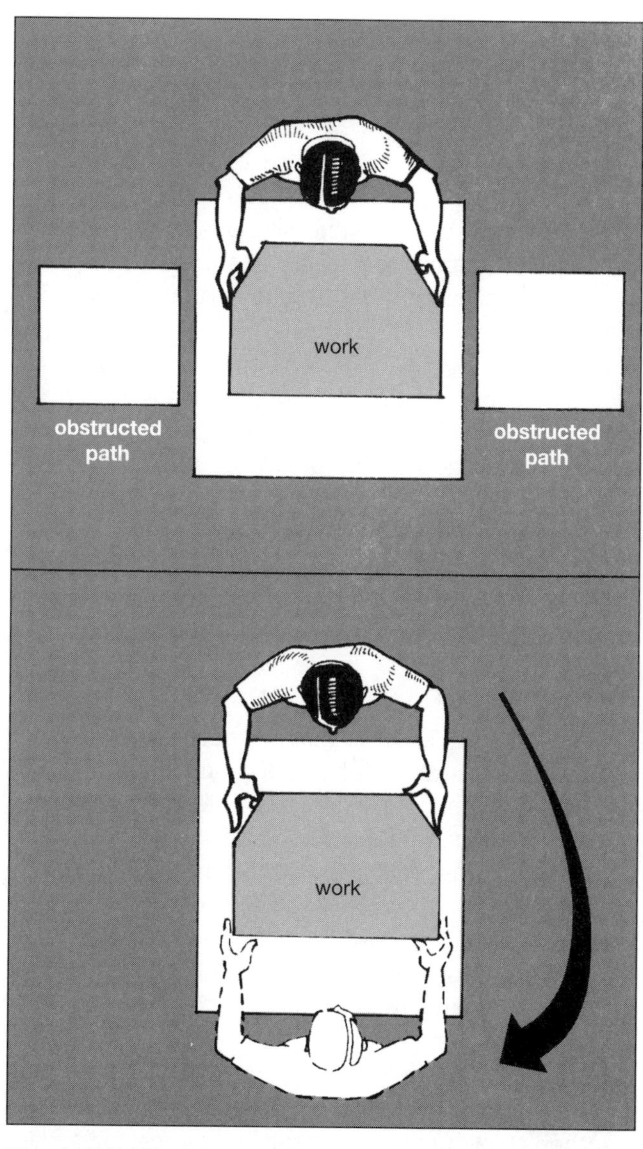

Figure 63b. If work is needed from both sides of the work item, arrange for free movements of the worker or for free rotation of the work item, so that the work is done in front of the worker

CHECKPOINT 64

Allow workers to alternate standing and sitting at work as much as possible.

WHY

Alternating standing and sitting is much better than keeping either posture for a long period of time. It is less stressful, reduces fatigue and improves morale.

Alternating standing and sitting means combining different tasks, thus facilitating communication and the acquisition of multiple skills.

Strictly machine-paced work requires keeping the same posture. This is tiring and tends to increase mistakes. By providing chances for occasional sitting or standing, the work becomes better organized.

HOW

1. Assign work tasks so that the worker can do these different tasks by alternating standing and sitting while at work. For example, preparation while standing and sitting, power tool work while standing, inspection and recording while sitting.

2. If the main tasks are done at standing workstations, then allow for occasional sitting (e.g. for watchkeeping, recording or at the end of a series of work tasks).

3. If the main tasks are done in a sitting posture, then opportunities should be provided for occasional standing, e.g for collecting materials from storage, communicating with other workers or monitoring work results, or after completing one or a few work cycles.

4. If appropriate, organize job rotation so that the same worker can go through different jobs alternating standing and sitting.

5. If alternating standing and sitting at work is not at all possible, insert short breaks to allow for the change.

SOME MORE HINTS

- If it seems difficult to introduce the new routine of alternating standing and sitting, just try to see if such changes are possible by providing standing workers with chairs for occasional sitting and by providing sitting workers with an additional space where some secondary tasks can be done while standing. This trial may facilitate a new routine.

- Multiple skills are increasingly important for various kinds of work. In arranging multi-skilled work to be done by a group of workers, it is possible to combine standing and sitting tasks for each individual worker.

POINTS TO REMEMBER

Assign work tasks in order to create opportunities to alternate standing and sitting for greater efficiency and comfort.

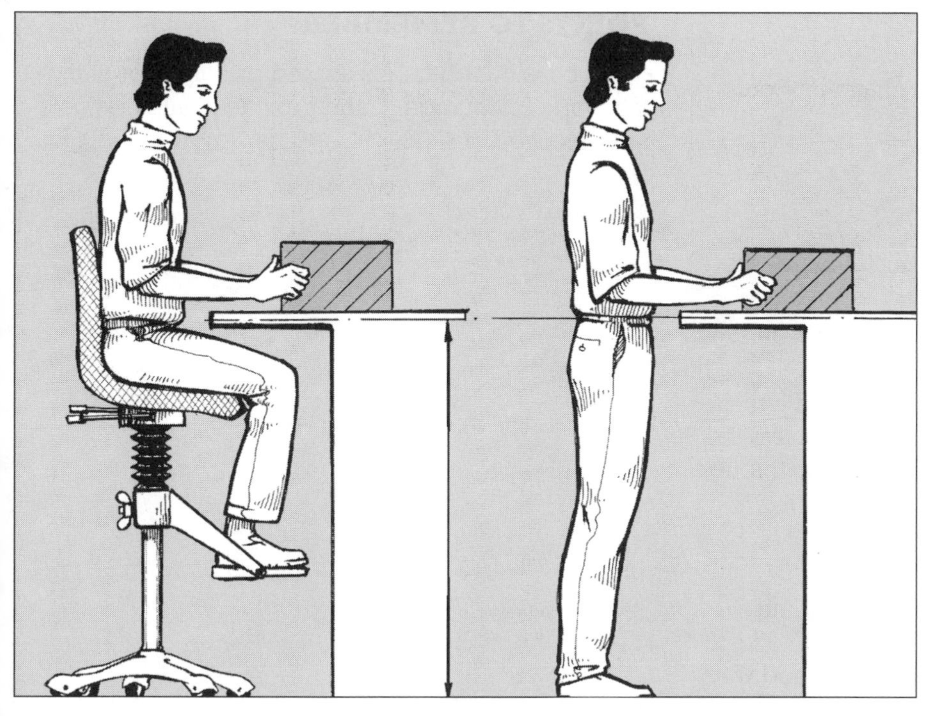

Figure 64a. For alternating standing and sitting postures when doing the same or similar tasks at a work table, a high stool with a good footrest is useful. Make sure that there is enough leg room in either posture

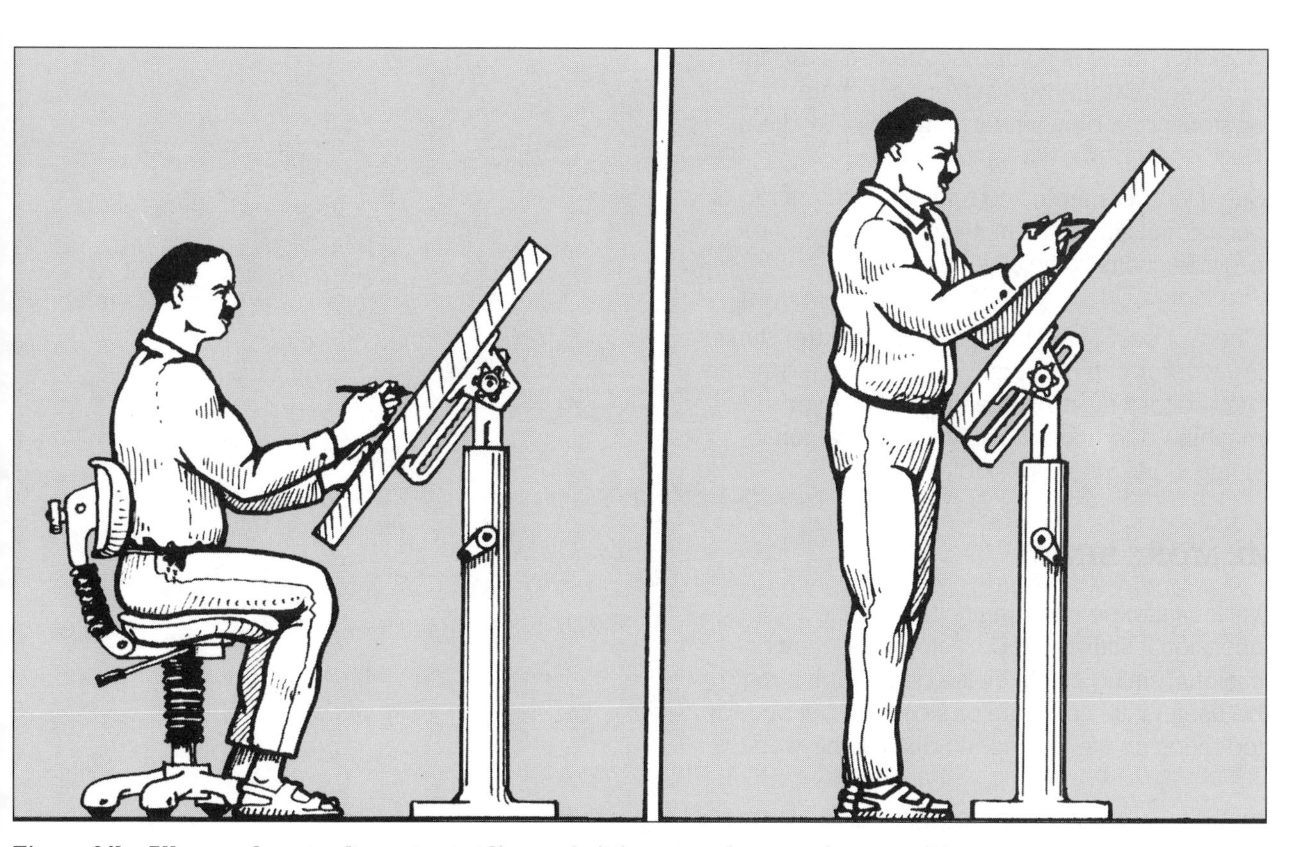

Figure 64b. Allow workers to alternate standing and sitting at work as much as possible

CHECKPOINT 65

Provide standing workers with chairs or stools for occasional sitting.

WHY

Standing all the time is very tiring. It increases pain in the back, legs and feet, and affects work quality. Occasional sitting helps to reduce fatigue.

Standing all the time is often considered a matter of discipline. But most standing workers do have the chance to sit and should be allowed to do so through the provision of chairs or stools. This helps to increase work quality and job satisfaction.

If some of the tasks carried out by standing workers can be done sitting, arrange for this to be done. Alternating sitting and standing is a good way of organizing work.

HOW

1. Provide a chair or stool near each standing worker. If there is no immediate space for this purpose near the workstation, put chairs or stools or a bench near a group of workers.

2. See if workers are using makeshift chairs for occasional sitting. Then it is certainly better to provide chairs "officially".

3. Check if part of the tasks assigned to the standing worker can be done while sitting (e.g. some preparatory tasks or keeping watch over the machine operation). Arrange for occasional sitting work, where possible.

SOME MORE HINTS

- Various inexpensive chairs can be used for occasional sitting. Support stools for easy occasional sitting can likewise be helpful. It may be useful if these chairs or stools do not occupy too much space and do not disturb the work.

- Make sure that the place for occasional sitting is accessible and safe.

POINTS TO REMEMBER

Occasional sitting is a good principle for standing work. Encourage it, and provide chairs or stools near the workstation.

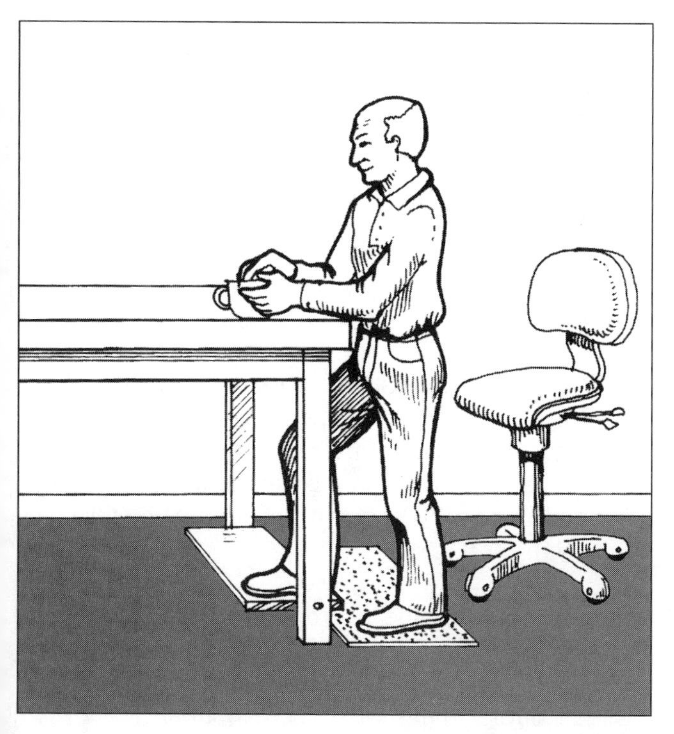

Figure 65a. Use a variety of means to ensure the comfort of a standing worker

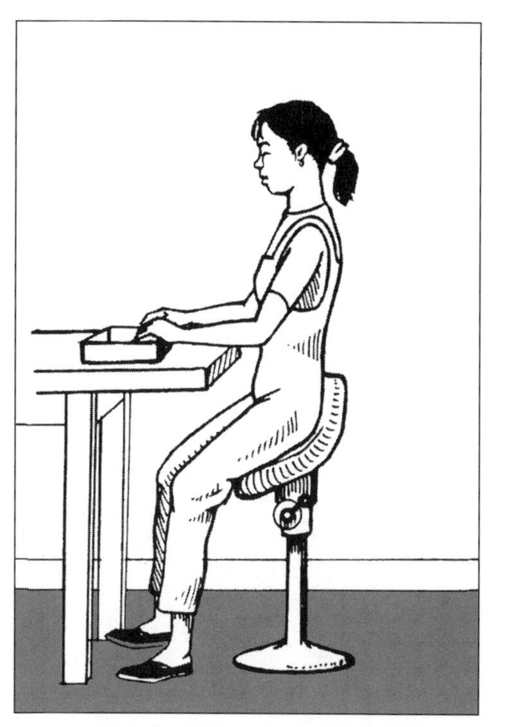

Figure 65b. Provide standing workers with chairs or stools so that they may sit occasionally

Figure 65c. At a workstation for standing work, a high chair for occasional sitting while watchkeeping can be useful

CHECKPOINT 66

Provide sitting workers with good adjustable chairs with a backrest.

WHY

Seated work seems comfortable compared with other forms of work. However, sitting for long hours is also tiring. Good chairs reduce fatigue, improve work efficiency and increase job satisfaction.

Often it is not considered worth while to invest money in chairs. But consider that a chair can last for years, and that the cost per day is only a very small fraction of the labour cost (an estimate is 0.1 per cent or even as low as 0.01 per cent). A good chair that improves productivity and job satisfaction more than offsets this minimal cost.

HOW

1. A suitable seat height is the height at which the worker can sit with the feet placed flat and comfortably on the floor and without any pressure to the back of the lower thigh. Provide a chair with adjustable height. Height adjustment should be very easy while sitting on the chair.

2. If an adjustable chair is not feasible, each worker should use a chair of correct height, or alternatively use a footrest or seat cushion in order to attain the correct floor/seat height difference.

3. Use a padded backrest that supports the lowest part of the back (often called the lumbar area) at waist level (about 15-20 cm above the seat surface) as people will lean both forward and backward in the chair. The backrest should also support the upper back for occasional leaning backward.

4. Provide a good seat surface with some padding, neither too hard nor too soft, so that the worker can easily change the sitting posture in the chair.

5. Ensure good mobility required for the work and for occasional changes of the sitting posture while in the chair. Five-leg chairs with castors are good for many seated tasks.

SOME MORE HINTS

- Ensure a good combination of correct seat height (lower end of kneecap level) and correct working height (elbow level). It is wrong to use a seat higher than the correct height in order to make the elbow level reach a high work table, because a high chair oppresses the thighs and restricts leg movements; this is very tiring for the worker.

- Do not use armrests for work that requires a lot of arm movements, as they inhibit mobility. Armrests are sometimes useful to give support for the whole arm (in this case, supporting the whole arm is better than just supporting the wrist).

- After adjusting the seat height so that the work is slightly below the elbow, the feet may dangle. This happens when the working height is not adjustable. In this case, use a footrest.

POINTS TO REMEMBER

Provide "ergonomic chairs" adjustable to each worker for correct seat height and with a good backrest. The chair should allow good mobility in the chair. Do not to forget to instruct all workers how to adjust their chairs.

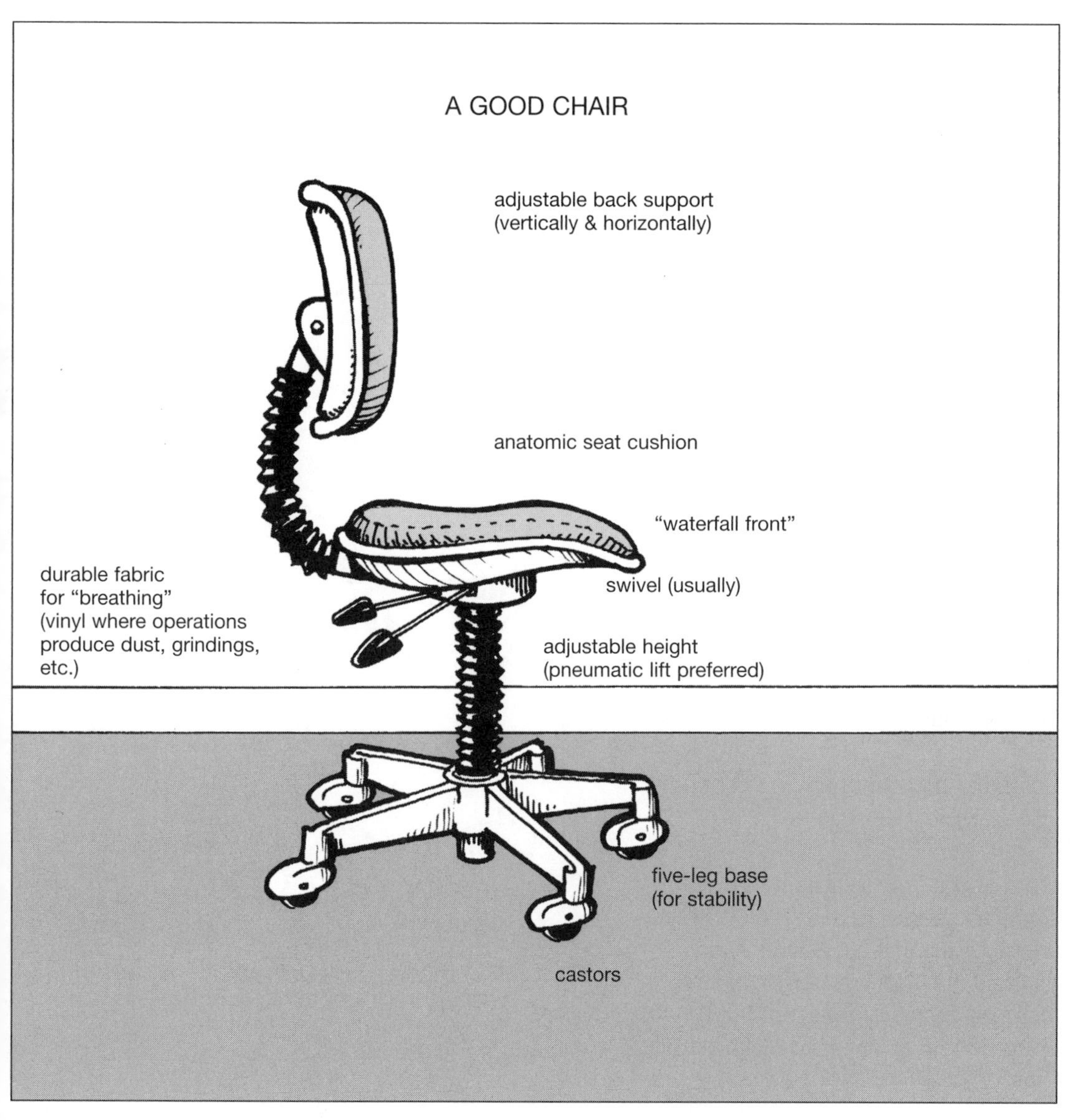

Figure 66. Provide sitting workers with good adjustable chairs with a backrest

CHECKPOINT 67

Provide adjustable work surfaces for workers who alternate work between small and large objects.

WHY

For convenient work operations, what matters most is easy-to-reach height and distance of the hand movements. Not only the work-table height but also the size of the work items handled must be taken into account. For different sizes of work item, the work-table height must be adjustable.

Adjustable work surfaces increase the efficiency of a worker dealing with objects of different sizes.

HOW

1. Provide an adjustable work surface that allows hand operations within the waist-heart level range for a standing worker (or within the elbow-heart level range for a sitting worker) in dealing with objects of different sizes.

2. Consider the maximum and minimum sizes of objects that the worker must deal with.

3. Instruct the workers about how to adjust the work-surface height according to the size of the object. If the work table itself is not adjustable, the adjustment of the work-surface height is still possible by placing platforms of different thicknesses or by stacking them in order to obtain the appropriate height.

4. If possible, reduce the number of changes in object size per work shift.

SOME MORE HINTS

- Make sure that all frequent operations are done at around elbow level. Avoid in any case working above shoulder level or below knee level.

- If a work surface is used by different workers, ensure that they can adjust their own workstation to their individual size and needs.

POINTS TO REMEMBER

Provide adjustable work surfaces to enable workers to perform work around elbow level for different sizes of work item.

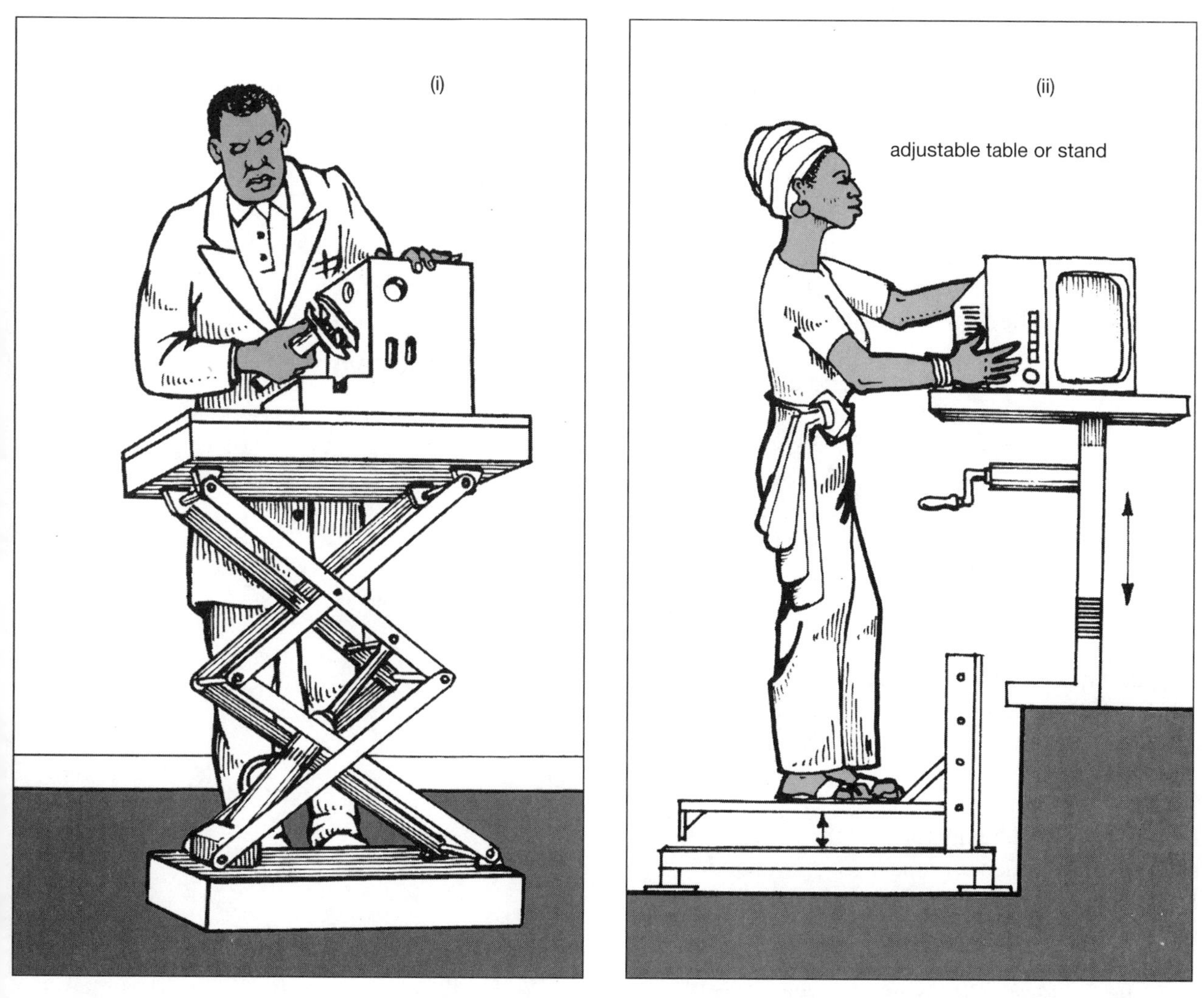

Figure 67. (i) and (ii) Provide adjustable work surfaces for workers who alternate work between small and large objects

CHECKPOINT 68

Use a display-and-keyboard workstation, such as a visual display unit (VDU), that workers can adjust.

WHY

Adjusting the workstation to preferred positions of displays and controls in relation to the worker's eye and hand levels can reduce strain (in particular neck, shoulder and back pains) and absenteeism.

If the display screen, keyboard and chair height can be easily changed, both larger and smaller workers can use the same VDU and work comfortably. This allows more efficient use of VDUs.

Adjustable workstations can increase productivity and reduce costly errors.

HOW

1. Provide chairs and tables whose height can be adjusted by workers. This allows comfortable sitting and the positioning of the keyboard at elbow level.

2. If a work table of adjustable height is not available, use low tables for smaller workers and higher tables for larger workers so that the keyboard is at elbow level.

3. If low tables for smaller workers are not available, use a platform to raise the chair so that the keyboard is at elbow level with both feet flat on the platform.

4. Raise or lower the existing table height, so that the keyboard is at elbow level while the worker is sitting comfortably.

5. Adjust the positions of the document holder, desk light or other items so that the worker can see and work easily and comfortably.

SOME MORE HINTS

- Always try to obtain adjustable tables in addition to adjustable chairs. Buy ones that workers can easily adjust themselves.

- Make sure that larger workers also have enough leg and knee room.

- Instruct all workers about how to adjust the workstation. Let workers adjust positions of the chair, and the tilt and height of the screen, keyboard and document holder according to their preferences, while taking into account the guidelines given here.

- It is advisable for workers to sit in different ways during work. Keeping one position for a long period is tiring.

POINTS TO REMEMBER

Adjusting display height, keyboard height and chair height in a good combination is the first step towards reducing VDU workers' complaints.

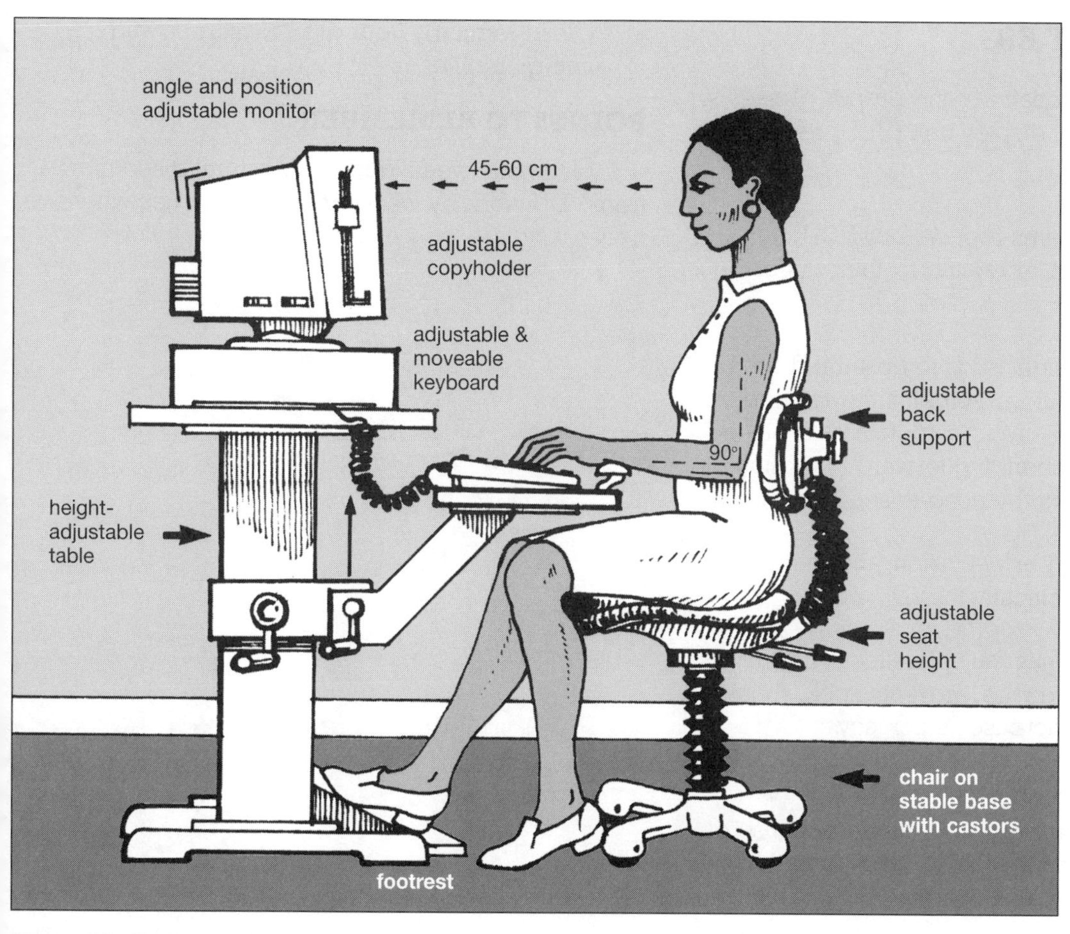

Figure 68. Each visual display unit user must be able to find his or her best work position

CHECKPOINT 69

Provide eye examinations and proper glasses for workers using a visual display unit (VDU) regularly.

WHY

Most visual problems reported by VDU workers are a result of, or related to, their uncorrected eyesight.

Uncorrected eyesight adds to postural discomfort and neck and shoulder complaints. Many operators adopt a poor posture to compensate for their visual difficulties (e.g. bending forward to reduce the viewing distance, tilting the head to see better).

Few people have perfect vision, and many need corrective lenses specifically for VDU work.

Corrected vision has multiple effects; it reduces visual fatigue and headache, prevents neck, shoulder and back pains, and increases efficiency.

HOW

1. Anyone experiencing visual fatigue because of display-and-keyboard (VDU) work should undergo a full eye examination.

2. Provide eyesight correction specifically suited to VDU work. Lenses for one task may not be suitable for another task.

3. Check VDU operators' eyesight as part of the regular health check-up programme, at least once every two years. Inform your optician (vision care specialist) about the different visual tasks that you perform.

SOME MORE HINTS

- Everybody knows that vision changes with age, accompanied by rapid reduction in visual performance. Yet there are many people who have not yet corrected their eyesight for their everyday work.

- Remember that requirements for corrective lenses for a visual display screen are different from those for reading a paper copy. A display screen requires a longer viewing distance (more than 50 cm) and a viewing angle that cannot be reached by bifocal lenses.

- Clean regularly your glasses, screen and anti-glare filter.

POINTS TO REMEMBER

Minimize eye fatigue and postural discomfort from VDU work by wearing proper glasses wherever necessary.

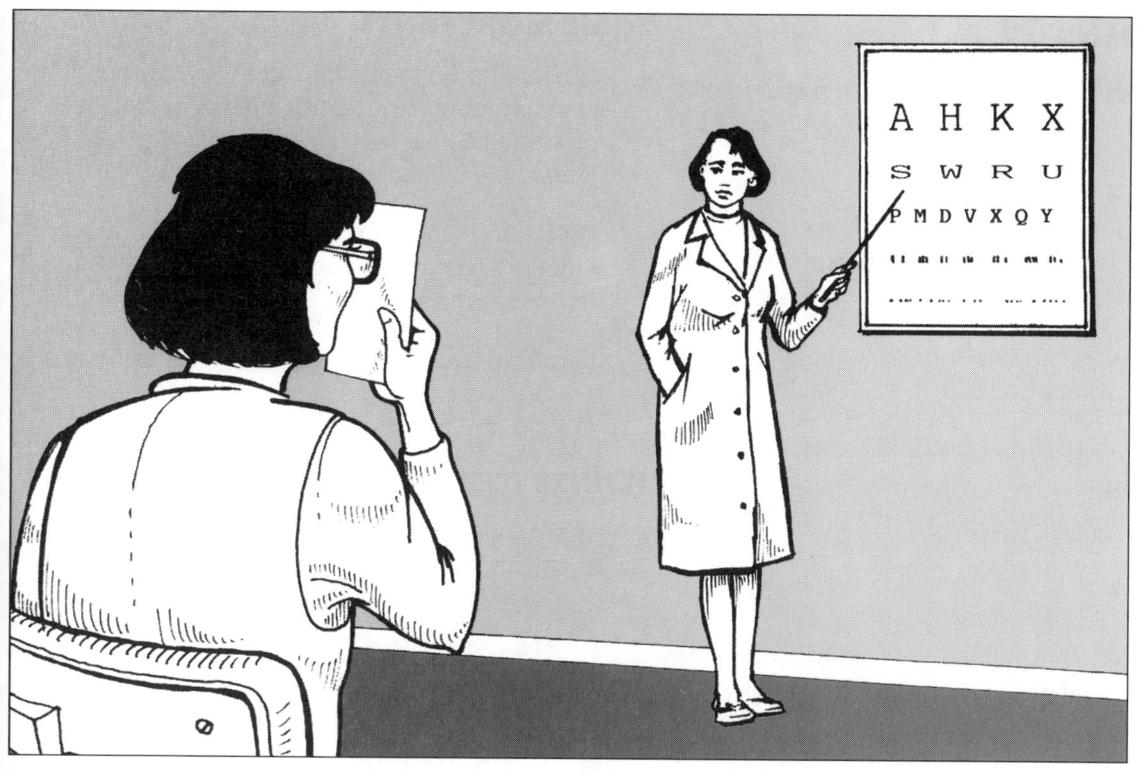

Figure 69a. Provide eye examinations for workers regularly using a visual display unit

Figure 69b. Provide eyesight correction specifically suited to visual display work

CHECKPOINT 70

Provide up-to-date training for visual display unit (VDU) workers.

WHY

VDU work is developing very fast. Up-to-date training ensures optimum utilization of computer facilities and resources.

Properly planned and up-to-date training improves VDU workers' comfort and satisfaction.

Training to bring workers up to date on available facilities and programmes minimizes costly errors and system breakdowns, and improves product quality.

HOW

1. Depending on individual skills, all VDU operators should undergo basic introductory training on:

    (a) the purposes and major functions of the system, and how system components work and are interconnected;

    (b) how to use different equipment and how to adjust a workstation (including display screen, keyboard and chair height, viewing angle and distance, contrast, lighting and glare prevention, and how to organize various items so that they are within easy reach).

2. Depending on individual needs, advanced training should be provided for VDU operators to acquire skills and knowledge relevant to their existing and future tasks on:

    (a) how to use and interact most effectively with available systems for performing the different tasks required;

    (b) what action to take in the case of system failures (including procedures to rectify such failures, disengaging from the system, whom to consult, etc.).

3. Make a survey of training needs and establish a training schedule. This schedule should be updated at regular intervals, say, every six months.

SOME MORE HINTS

- When introducing new programmes, procedures or equipment, organize training sessions for VDU operators. These sessions must allow enough time for individual exercises.

- Tailor the training programme according to individual needs and capabilities. Allow more time for those who need it.

- Organize a separate training course for new employees and make sure that they receive up-to-date training.

POINTS TO REMEMBER

Up-to-date training is the most effective way to utilize rapidly developing technologies. Such training ensures high work quality for every worker.

Figure 70. Up-to-date training through learning by doing improves the efficiency, comfort and satisfaction of visual display unit operators

CHECKPOINT 71

Involve workers in the improved design of their own workstation.

WHY

No one knows more about the job than the person who does that work every day. This worker is the best source of information on ways to improve equipment and productivity.

People are more likely to follow up their own ideas in improving their workstation. Involving workers in designing improvements ensures their cooperation in making full use of the modifications achieved.

HOW

1. Ask workers about any problems their work-station causes. Then get their ideas on how to overcome these problems.

2. Use good examples implemented locally as guide materials for formulating feasible improvements.

3. Discuss workers' suggestions immediately. Try to find suggestions which can be implemented right away or after a short period. If there are suggestions which cannot be accepted for technical, financial or other reasons, explain the reasons, restate the problem and ask for other suggestions.

4. Give people recognition for their ideas for improving their workstation. This will encourage future improvements.

SOME MORE HINTS

- Create a concrete opportunity for workers to propose ideas for improving their workstation. One good way is to set a deadline for ideas. Make it clear that these ideas will be discussed in order to find the most feasible solutions. People may not be accustomed to expressing their ideas and may not have enough practice to find realistic solutions; therefore provide enough time for them to think about the problem and propose a solution.

- Always use group discussion for studying the proposals, comparing options and identifying a feasible solution.

- Pick up multiple aspects of the workstation design at the same time. This stimulates various possible ideas and makes it easy to identify feasible improvements.

POINTS TO REMEMBER

Your best source of ideas for improving your workstation is the people who do the work every day.

Figure 71a. Discuss workers' suggestions about improving their own workplaces and workstations

Figure 71b. Formulate suggestions in a way that can be accepted by the staff involved. Issue information on the results of safety inspection rounds and proposed measures

Lighting

CHECKPOINT 72

Increase the use of daylight.

WHY

Daylight is the best and cheapest source of illumination. The use of daylight reduces energy costs.

The distribution of light in the workplace can be improved by using more daylight. Measures to use daylight are effective for years to come and greatly help to improve the efficiency and comfort of workers.

Using daylight is an environmentally friendly action.

HOW

1. Clean windows and remove obstacles that prevent the entrance of daylight.

2. Change the place of work or the location of machines so that the worker has more daylight.

3. Expand the size of windows or have windows placed higher to take advantage of more daylight.

4. Separate switches for different electric lights or for different rows of lights so that parts of lighting can be turned off when there is enough daylight at workplaces near windows.

5. Install skylights with semi-transparent material at proper intervals. Skylights can be installed in the existing roof by simple replacement of a few roof panels with translucent plastic panels.

SOME MORE HINTS

- Combine daylight with artificial lights to improve your workplace lighting.

- Be careful, as windows and skylights provide heat in hot weather (and cause heat loss in cold weather).

- In a hot climate, orient windows and openings away from direct sun heat or protect them from direct sunlight.

POINTS TO REMEMBER

The use of daylight reduces your electricity bills and is environmentally friendly.

Figure 72. Using daylight reduces energy costs

CHECKPOINT 73

Use light colours for walls and ceilings when more light is needed.

WHY

The choice of colours for walls and ceilings is of great importance, because different colours have different reflectivity. White has the highest reflectivity (as high as 90 per cent), while dark colours have much lower percentages.

Light-coloured walls and ceilings are energy saving as they produce higher room illumination with fewer lights.

Light-coloured walls and ceilings make rooms more comfortable. This helps create an environment conducive to efficient work.

For precision and inspection tasks requiring accurate colour recognition, light-coloured surfaces are essential.

HOW

1. To provide adequate reflection of light, use a very light colour for ceilings, such as white (80-90 per cent reflectivity) and a pale tint for walls (50-85 per cent).

2. Avoid large differences in the brightness of walls and ceilings.

3. Do not use glossy or shiny materials or paints for the finish of these surfaces, in order to prevent indirect glare.

4. Use a combination of a white ceiling and lighting units with upward openings so that the ceiling reflects light from the units and the lighting units reflect light from the ceiling. This produces well-distributed general lighting.

SOME MORE HINTS

- Clean the walls and ceilings regularly as dust and dirt absorb a large proportion of light.

- Openings in the tops of lighting units not only allow ceiling illumination but produce better light distribution and lower dirt accumulation than closed-top units.

POINTS TO REMEMBER

Light-coloured walls and ceilings create a comfortable and effective working environment.

Figure 73. (i) and (ii) Light colours for walls and ceilings improve lighting conditions and the workplace atmosphere

CHECKPOINT 74

Light up corridors, staircases, ramps and other areas where people may be.

WHY

Poorly lit or dark places cause accidents, especially when materials are being moved.

Staircases, back doors and storage rooms tend to be poorly lit and often become dumping sites. Often daylight does not reach staircases. Special attention to these areas is necessary.

Sufficient lighting in these areas can prevent damage to materials and products.

HOW

1. Clean windows and existing lights (lamps, fixtures, reflectors, covers) and change worn-out bulbs and tubes on staircases and ramps, and in corridors, storage rooms and other back-way areas.

2. Remove obstacles that prevent good distribution of light.

3. Relocate existing lights for better illumination of these areas. Add new lights after consulting workers.

4. Make best use of daylight by keeping some doors open or installing new windows or skylights.

5. Provide easy-to-reach electrical switches near the entrances and exits of corridors and staircases.

6. Paint surfaces with bright colours to make stairs and height differences clearly visible.

SOME MORE HINTS

- Lighting should be an important part of regular inspection and maintenance programmes.

- The level of illumination of staircases, corridors and storage areas may be lower than in production areas, but should be sufficient for safe moving and transport.

- Avoid automatic electrical switches if staircases, etc., are regularly used or if sudden shut-off may cause an accident.

POINTS TO REMEMBER

Better lighting on staircases and in corridors prevents accidents to workers and visitors, reduces product damage and improves the image of your enterprise.

Figure 74a. Good lighting on staircases and in corridors prevents accidents and reduces product damage

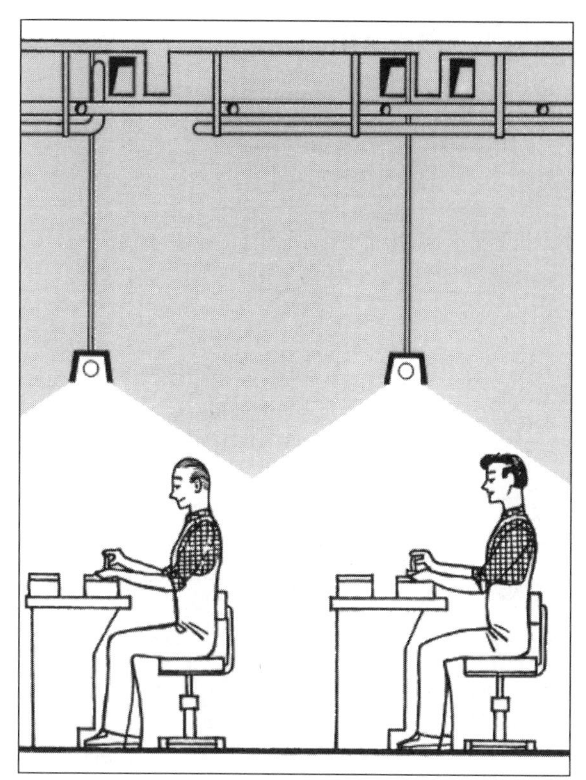

Figure 74b. Direct light

Figure 74c. Sharp shadows make it difficult to work

CHECKPOINT 75

Light up the work area evenly to minimize changes in brightness.

WHY

Changing the view from a bright area to a dark area demands adaptation of the eyes, which takes time and is tiring.

It is more efficient and more comfortable to work in a room which does not have large variations in brightness.

Suppression of flickering light is important. Such lighting is annoying to everybody and causes eye fatigue. It even produces some odd effects dangerous to sufferers of epilepsy.

Sharp shadows on the work surface are a source of poor work quality, low productivity, eye strain, fatigue and sometimes accidents. Eliminate shadows.

HOW

1. Eliminate isolated pools of bright light. These are uneconomic and disturb the even illumination of the workplace.

2. Consider whether changing the height or positions of some existing lights can improve lighting so as to create more even illumination of the workplace. Consider whether adding some lights for general lighting can improve illumination.

3. At the same time as using daylight, light up workplaces away from windows if appropriate. For example, provide different switches for lights near windows and for lights illuminating places away from windows. In this way, the lights near windows can be turned off when there is enough daylight.

4. Eliminate shadow zones by having good distribution of lights and reflection from walls and ceilings, as well as by better layout of workstations.

5. Suppress flickering light by changing worn-out fluorescent lamps. If necessary, use filament bulbs instead.

SOME MORE HINTS

- It is important not always to rely on installing electric lighting. A good combination of different means of improving lighting will help you a great deal. Use daylight properly. Use reflection from walls and ceilings. Combine general and local lights. Improve the layout of workstations.

- To attain even illumination, a mixture of direct and reflected light provides the best visibility. Do your light units have upward openings?

- In order to avoid the flickering (stroboscopic) effect of fluorescent lamps, which disturbs workers, consider the use of high-frequency fluorescent tubes or three-phase current connected to different fluorescent tubes. If nothing helps, try to cover the two ends of the tubes for about 10 cm at each end to mask the end flickers.

- For general lighting, it is often true that the higher the lights, the better the uniformity and dispersion of light.

POINTS TO REMEMBER

Avoid large differences in brightness in the workplace due to uneven distribution of bright lights and lack of adequate reflection.

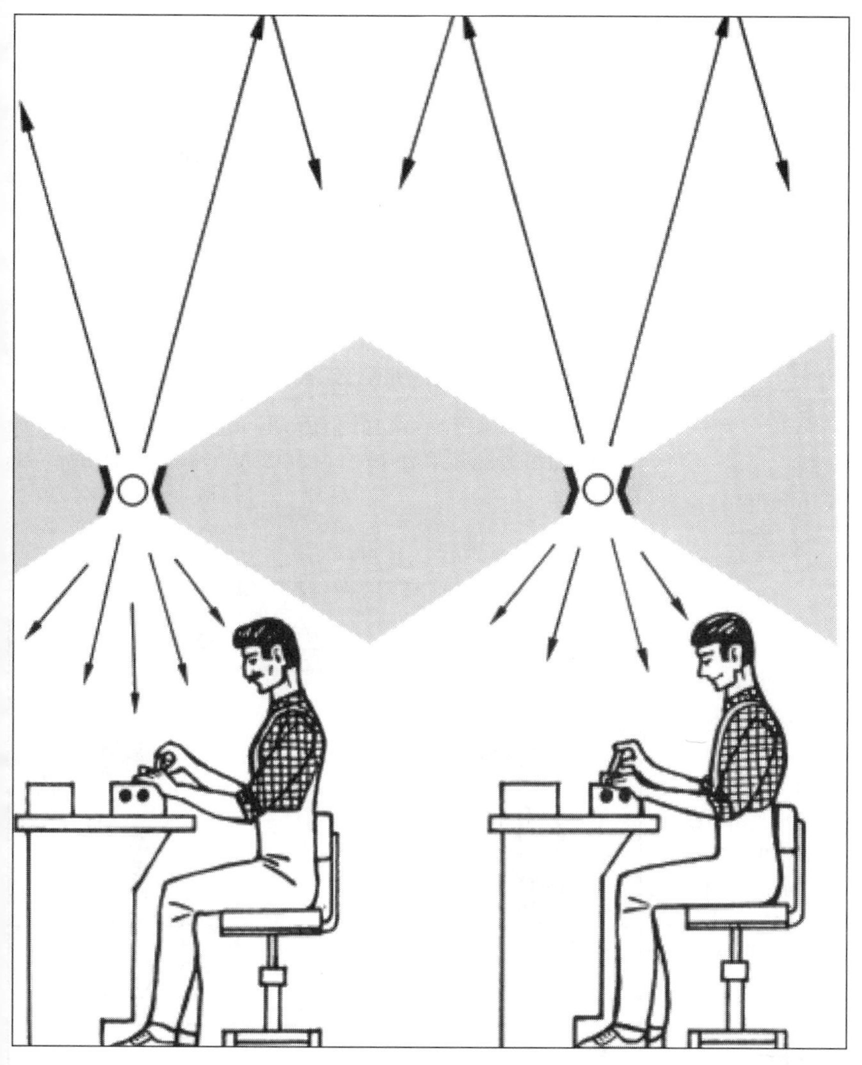

Figure 75. A mixture of direct and reflected light provides the best visibility

CHECKPOINT 76

Provide sufficient lighting for workers so that they can work efficiently and comfortably at all times.

WHY

Sufficient lighting improves workers' comfort and performance, making the workplace a pleasant place to work.

Sufficient lighting reduces work errors. It also helps to reduce the risk of accidents.

Adequate and good-quality lighting helps the workers to see the work item quickly and in sufficient detail as the tasks require.

HOW

1. Combine the use of natural light (through windows and skylights) and artificial lighting (electric lights), as this is usually the most pleasant and cost-effective.

2. Provide sufficient lighting considering the nature of the tasks performed at various workstations. For example, more light is needed for precision work (seeing smaller objects) and for materials having lower reflective properties (e.g. work with dark-coloured cloth).

3. If appropriate, change the positions of lamps and the direction of light falling on objects. You can also try to change the positions of workstations to obtain better lighting from existing lamps.

4. Consider the age of your operators. Older workers need more light. For example, an operator aged 60 needs five times more light to comfortably read a printed text than a 20-year-old.

5. The level of lighting also depends on the time available for seeing objects. The faster the task (e.g. identifying defects on passing objects), the more and better arrangements of light are required.

SOME MORE HINTS

- Regularly maintain the existing lighting. Clean lamps, fixtures and reflectors, as well as windows, ceilings, walls and other interior surfaces. Change worn-out bulbs and tubes.

- Light-coloured walls reflect more light and provide better lighting conditions and a good workplace atmosphere.

- Most people over the age of 40 need glasses. Regular vision checks are recommended as part of the workers' health programme.

POINTS TO REMEMBER

Provide sufficient and good-quality lighting at minimum cost. There are various ways to improve lighting.

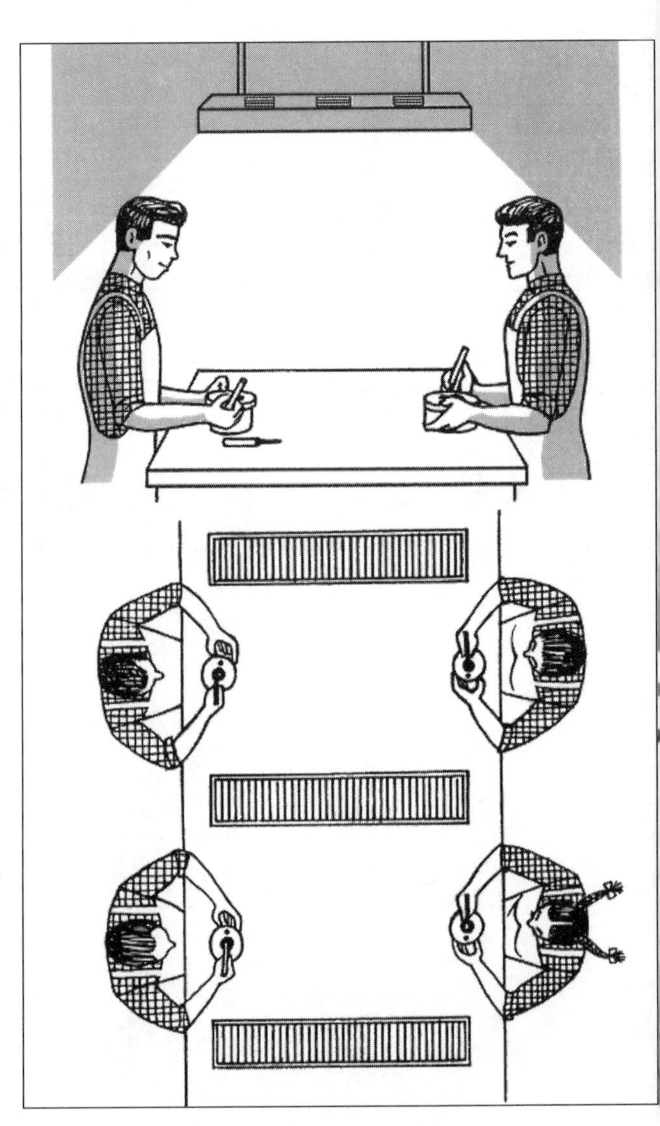

Figure 76a. Bench lighting for work with larger objects

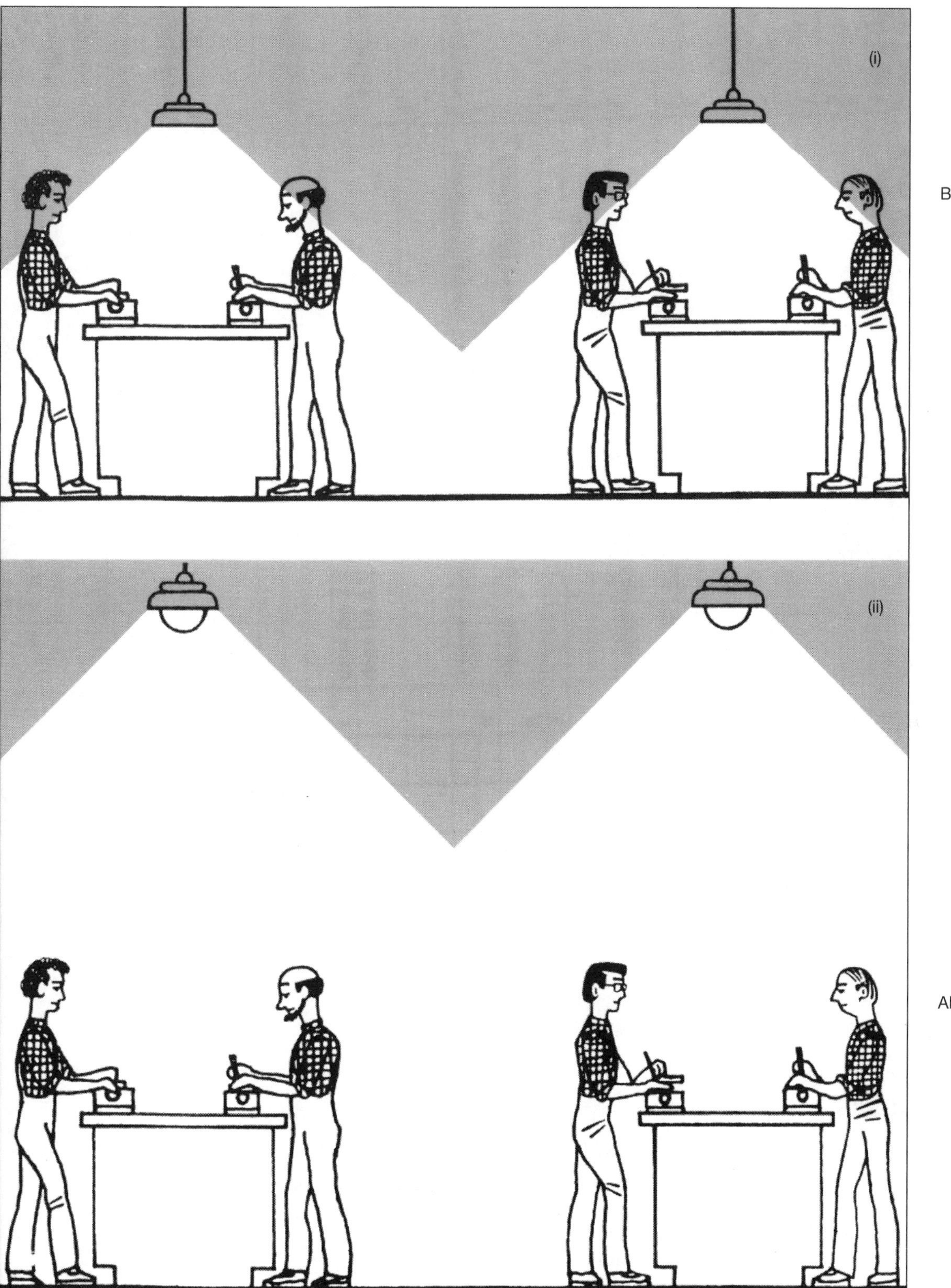

Figure 76b. (i) and (ii) Higher lighting gives better dispersion

CHECKPOINT 77

Provide local lights for precision or inspection work.

WHY

More light is required for precision or inspection work than for normal production or office work.

Appropriately placed local lights greatly improve the safety and efficiency of precision or inspection work.

A combination of general and local lights helps to meet the specific requirements of different jobs and helps to prevent disturbing shadows.

HOW

1. Place local lights near and above precision work and inspection work. The local lights having a proper shield should be in a position where they will create neither glare for the worker nor disturbing shadows. No naked bulb should be used as a local light.

2. Where appropriate, use local lights that are easy to move and arrange in the desired positions.

3. Use local lights that are easy to clean and easy to maintain.

4. Use daylight bulbs or tubes (white fluorescent bulbs or tubes) for colour recognition tasks.

5. Always ensure a good combination of general and local lights so that each workstation has an appropriate contrast between the work-point and the background.

SOME MORE HINTS

- Make sure that local lighting does not restrict the operator's view.

- When using a local light, mount it on a rigid isolated structure instead of on vibrating machines.

- Use deep lampshades for local lights and paint the inner edge of the shades in a dark matt colour to prevent bright reflections.

- Local lights with filament tubes produce heat, often resulting in discomfort for the worker. Use fluorescent lamps instead. Various kinds of fluorescent lamp are available.

POINTS TO REMEMBER

Local lights, properly placed, reduce energy costs and are surprisingly effective.

Figure 77a. A combination of general and local lights helps to meet the specific requirements of different jobs

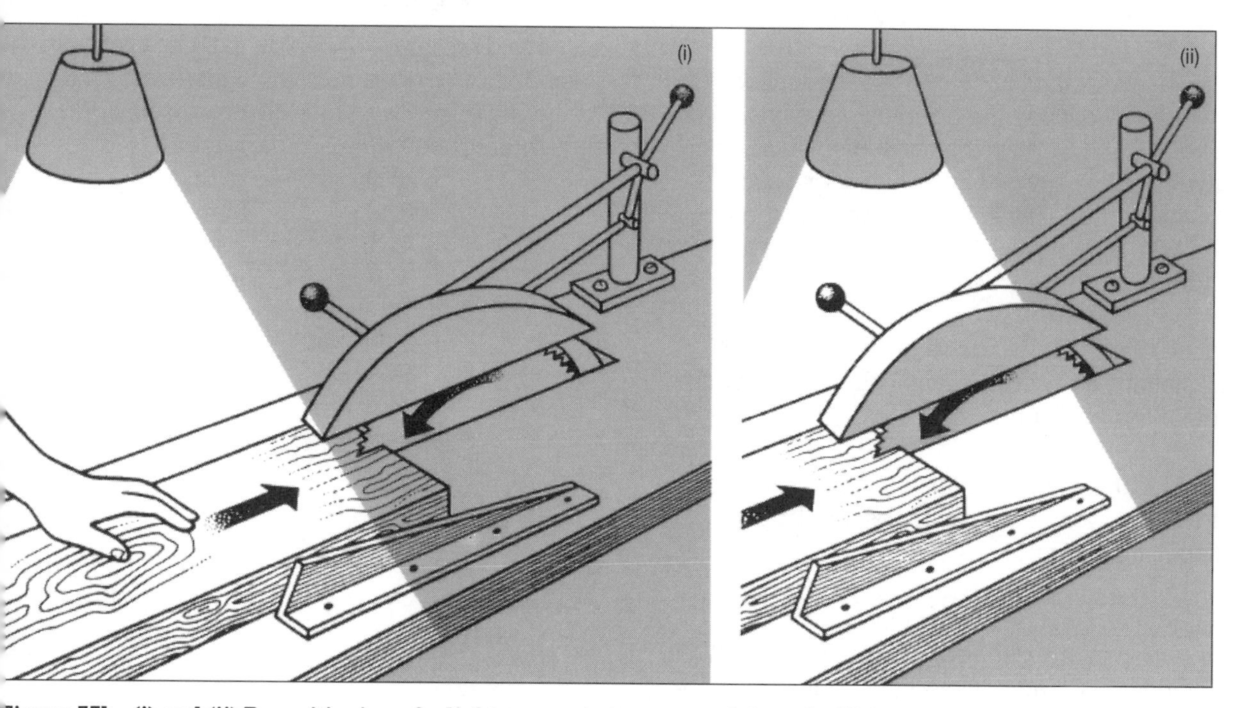

Figure 77b. (i) and (ii) Repositioning of a light source to improve safety and efficiency

CHECKPOINT 78

Relocate light sources or provide shields to eliminate direct glare.

WHY

Direct or reflected glare greatly reduces everyone's ability to see. For example, the image of bright overhead lights reflected in a display screen disturbs screen reading.

Glare at work is a cause of discomfort, annoyance and eye fatigue. Glare also reduces the worker's performance, causing low work quality and low productivity.

There are means to eliminate glare. Eliminating direct glare from windows or light sources, as well as indirect glare from reflections, reduces workers' complaints and allows for more efficient use of machines.

HOW

1. Place display panels or screens in a place not facing the window and not with the window directly behind them in order to reduce direct glare from daylight.

2. Do not position any naked light bulbs or tubes within the field of vision of workstations.

3. Relocate overhead lamps or raise them so that they are well outside the normal field of vision of workers. Place workstations with display screens between rows of overhead lights so that such lights are not directly above the workstation and so that the operator's line of sight is parallel to the overhead lights.

4. Reduce glare from windows or neighbouring workstations by using curtains, blinds, partitions or desk-top partitions.

5. Mount local lights (task lights) low enough and shade them well to hide all bulbs and bright surfaces from the normal field of vision.

6. Change the direction of light coming to the workstation in order to avoid glare, e.g. so that the workers, instead of facing the light source, now have their sides or backs towards it.

SOME MORE HINTS

- Change the window glass to translucent glass instead of transparent glass.

- For local lights placed near the point of work, use deep lampshades and paint the inner edge of the shades in a dark matt colour. If appropriate, use shields between the lights and the eyes or between the lights and the display screen.

- Eliminate distracting reflection and glare by reducing overhead lighting (turning off some lights, where possible, and providing desk lamps for workers), drawing curtains or adjusting the blinds. Naturally make sure that there is sufficient light for performing the various tasks required.

- Use medium-reflectance and low-contrast walls, ceilings and floor surfaces (do not use colours that are too bright or of high contrast vis-à-vis a VDU screen, or very dark and gloomy colours).

- Use indirect upward lighting to distribute the light over a greater area of the ceiling. The reflected light coming from the ceiling helps to eliminate over-bright spots and to minimize glare.

POINTS TO REMEMBER

By avoiding direct glare you can greatly improve the visibility of work items without increasing light intensity. Glare-free work greatly enhances work quality and reduces workers' discomfort.

Figure 78a. A shaded lamp should be placed at the appropriate height

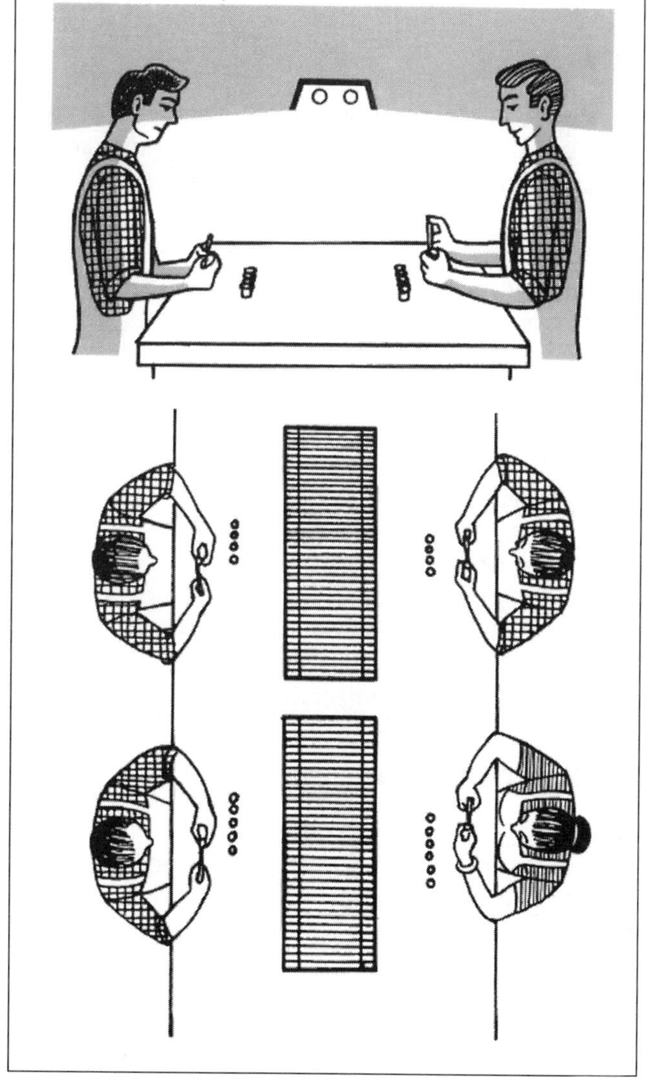

Figure 78b. Bench lighting for work with small objects

CHECKPOINT 79

Remove shiny surfaces from the worker's field of vision to eliminate indirect glare.

WHY

Indirect glare (reflected glare), like direct glare, causes distractions and reduces the ability to see.

The discomfort and annoyance of indirect glare continue whenever work is done, thus leading to eye fatigue and low performance.

HOW

1. Reduce reflection from polished surfaces or glass surfaces of equipment by changing the position of the light source.

2. For surfaces producing disturbing reflections, use matt surfaces instead of glossy, shiny or bright colours. Desks and work surfaces should be matt.

3. If appropriate, lower the brightness of the light source while making sure that there is still sufficient light available for performing the task comfortably and efficiently.

4. Make the immediate background brighter by placing a light-coloured surface behind the task area. The working field should be brighter in the middle and darker towards the edge.

SOME MORE HINTS

- Older workers are more sensitive to glare. They need better lighting arrangements and totally glare-free conditions.

- Try different positioning of the workstation, work objects and workstation lighting to find the best arrangements and glare-free conditions.

POINTS TO REMEMBER

Light reflection from shiny surfaces causes eye fatigue and low performance. Remove disturbing reflections from the field of vision at work.

Figure 79a. Reflected (indirect) glare

Figure 79b. Indirect glare, where the light is being reflected

Figure 79c. Glare reflected from a polished surface reduces visibility

CHECKPOINT 80

Choose an appropriate visual task background for tasks requiring close, continuous attention.

WHY

Visual tasks that demand close, continuous attention are performed with much less strain if their background is free from eye-catching distractions.

When the workpiece is small and held close to the eyes, a plain background without disturbance to the eyes is particularly important for high-quality work.

Workers doing critical assembly or precision work may be seriously distracted by neighbouring operations, such as moving machines or machine parts, or hand movements of a second worker sitting opposite. Simple measures can prevent such distractions.

HOW

1. Place a screen giving a plain visual background behind the point of operation so as to shield eye-catching distractions.

2. Place a partition between neighbouring work-stations when the operations at one workstation distract workers carrying out operations at the other. Similarly, place a desk-top partition between workers doing operations at the same workbench or work table.

3. If necessary to see clearly the outlines of small, flat objects, use a sheet of light-diffusing glass or plastic which is lit from behind by lamps or reflectors.

4. Avoid shadows thrown on the object against a brighter background; relocate the light source so that the light comes from above or from over the shoulder of the worker instead of coming from behind the visual task.

SOME MORE HINTS

— An appropriate jig on which a work item is placed with clear separation from other things often helps obtain clear vision of the work item.

– Avoid prolonged work in an isolated pool of light in the middle of a darkened interior. In such a case, the eyes become fatigued as they have to readjust every time the worker looks away from the brightly lit work-point.

– Partitions placed between neighbouring work-stations and neighbouring workers should not disturb communication between workers. For example, desk-top partitions should be low enough to allow some visual and verbal contact between workers.

POINTS TO REMEMBER

The elimination of potential distractions from the visual task background contributes very much to efficiency and safety.

Figure 80a. A low partition helps to avoid visual distraction

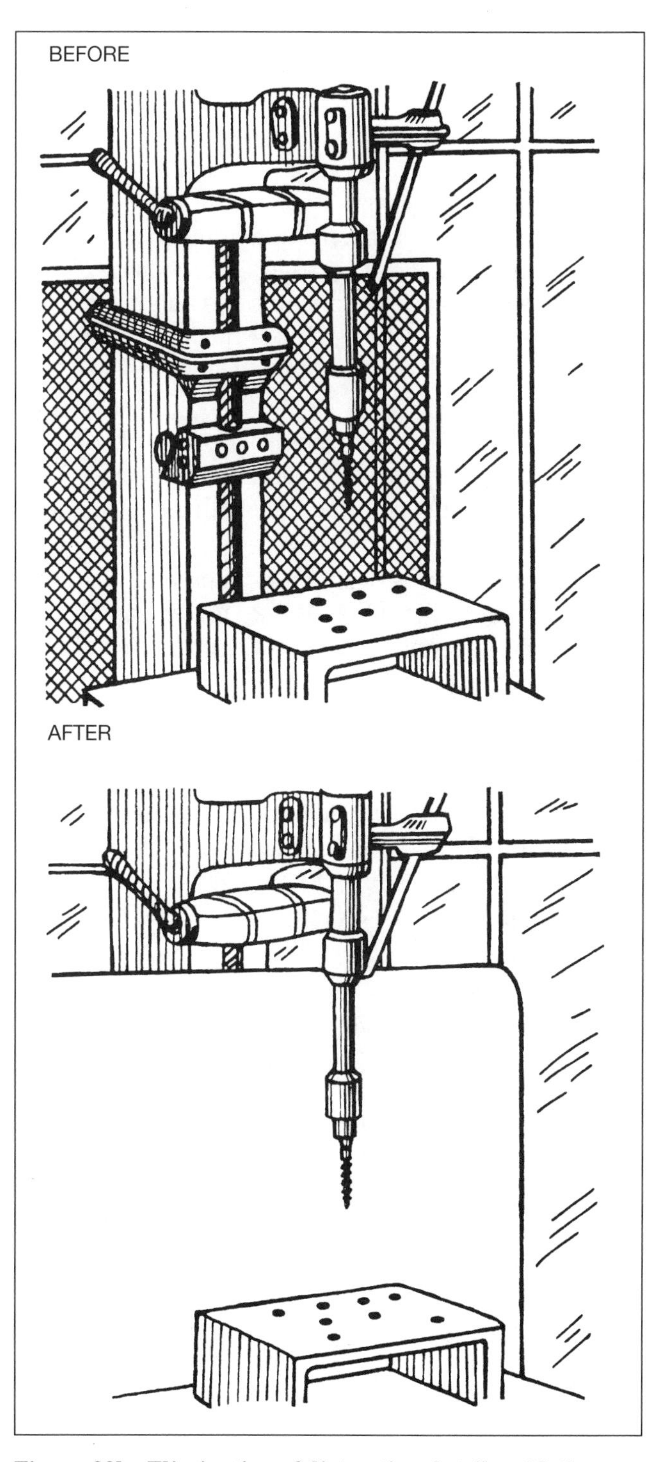

Figure 80b. Elimination of distracting details with the help of a screen

CHECKPOINT 81

Clean windows and maintain light sources.

WHY

Poorly cleaned or maintained light sources may reduce lighting annually by a large percentage. Gradual deterioration of lighting goes unnoticed, making it a hidden source of poor-quality work and accidents.

Well-maintained and clean windows and light units reduce energy consumption by increasing daylight and by producing more light with less wattage.

The maintenance of light units increases the service life of bulbs and tubes. Timely replacement of worn-out fluorescent tubes prevents problems from flickering lights.

HOW

1. Make cleaning of windows and light units a regular part of the weekly routine. Assign cleaning responsibility to a trained person who also understands the danger of electric shocks.

2. Ensure that the maintenance personnel have at their disposal appropriate cleaning tools and ladders to reach the lighting units and windows.

3. Incorporate the replacement of worn-out bulbs and tubes as part of this maintenance programme. Encourage workers to report lighting problems and worn-out bulbs and tubes to the persons in charge of lighting maintenance.

4. As much as possible, use light units which have openings at the top as these openings allow an upward warm air-stream that helps to keep the reflectors clean.

SOME MORE HINTS

- When cleaning, use mild and non-abrasive detergents to avoid corrosion of the reflectors. Use water sparsely to make cleaning effective and to minimize the risk of accidents involving electricity.

- Note that the service life of various types of lamp varies widely. Ordinary filament bulbs may last approximately 700 hours; a fluorescent tube may last ten times longer.

- If the electrical lights are numerous, it may be economical to change all bulbs or tubes at fixed time periods. The leftover bulbs or tubes that are still working can be used for individual replacements before the predetermined point of time.

- Changing worn-out, corroded or stained reflectors is more economical than replacing the whole light unit. Aluminium foil is a good material or replacement for a reflector.

POINTS TO REMEMBER

Start a cleaning and maintenance programme for light sources and windows by designating persons responsible for it.

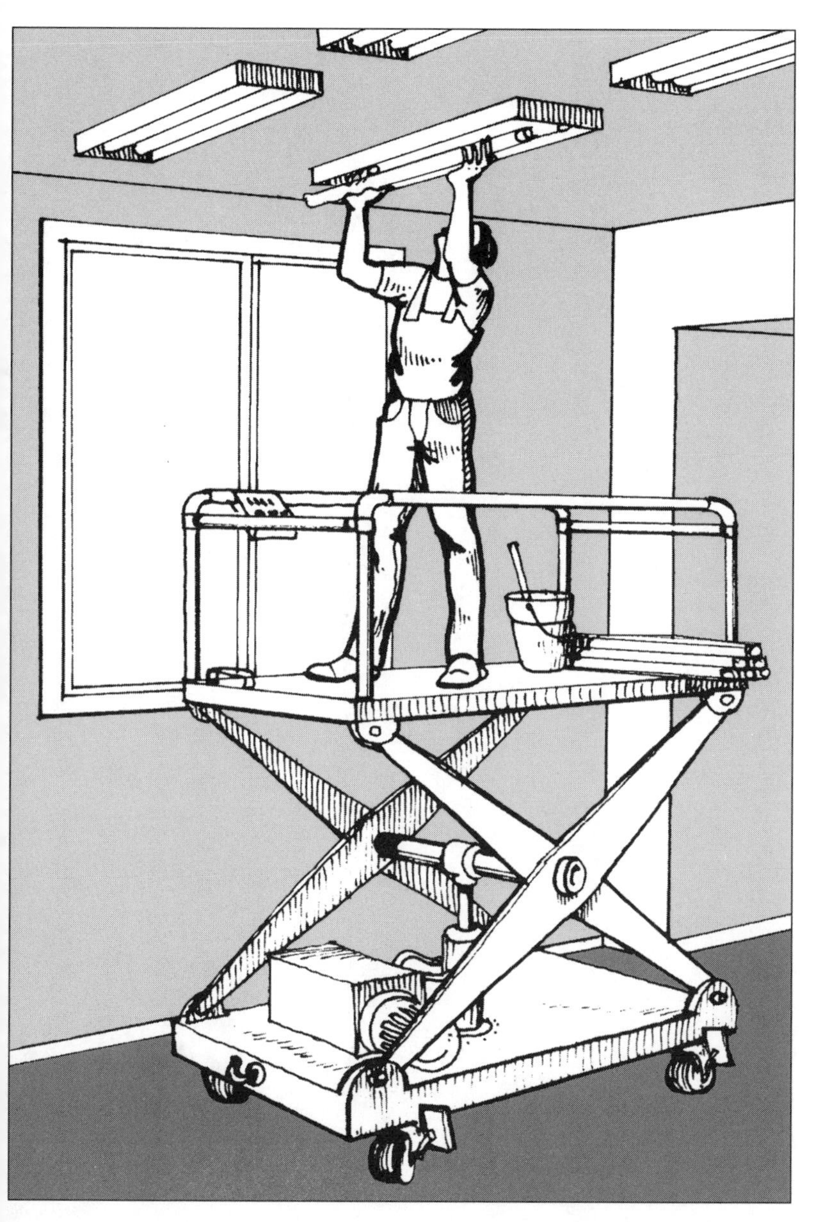

Figure 81. Clean windows and well-maintained light sources help to increase lighting

Premises

CHECKPOINT 82

Protect the worker from excessive heat.

WHY

Excessive heat can strongly influence working capacity. It may greatly decrease productivity and may increase errors and accidents.

Heat stress increases fatigue and may lead to heat-induced illnesses.

It is often difficult to control the workplace temperature. Hot process areas may be an unavoidable part of production. Regulation of temperatures may be impractical on production sites in a tropical climate, especially when the workroom air is polluted with dust or chemicals arising from production. In such a case, it is important to provide available means of protection against excessive exposure to heat.

HOW

1. Try multiple available measures to bring down the workplace air temperature. This is important when air-conditioning is not possible. Measures should include protection from outside heat coming into the workplace (solar heat), increased natural ventilation, isolation from hot machines and processes, and the provision of local exhaust systems for heated and polluted air.

2. Protect workers from heat radiation from hot machines and equipment and from hot surfaces (e.g. heated roofs or walls). The best way to reduce radiation reaching the workers is to put screens or barriers between the radiation source and the body. Also use insulated ceilings and walls. Where exposure to excessive heat sources is unavoidable, minimize the exposure time and wear protective clothing that can protect workers from heat radiation.

3. Avoid physically heavy work for workers who are simultaneously exposed to high temperatures or to strong heat radiation. Mechanize such work, or introduce rotation of workers so that the duration of exposure to excessive heat per worker is reduced.

4. Increase air velocity around the work area by means of fans and ventilators.

5. If possible, construct, inside the workplace, a small air-conditioned operation booth or room so that the operators can stay in it for most of their working time.

6. Minimize the period of time during which the workers are exposed to high temperatures or to strong heat radiation (e.g. by approaching an area where this is the case only if absolutely necessary or creating a work area behind a heat barrier for work that the workers can carry out without being exposed to strong heat radiation, a resting corner with good natural ventilation or fans, rotation schemes or frequent breaks).

SOME MORE HINTS

- It is often necessary or useful to combine the measures mentioned above. Avoid long hours of hot work and ensure sufficient breaks, especially during physically heavy work.

- Check if excessive heat is causing problems for product quality or for workers' health (e.g. by comparing work results between hot and cooler months or by interviewing workers and supervisors).

- Ensure a supply of cold drinking-water or beverages near the workplace. Liquids should be taken in small quantities and often.

- Provide good washing facilities and access to laundering of work clothes.

POINTS TO REMEMBER

There are multiple ways to reduce exposure to excessive heat. Do not give up even if air-conditioning is not possible. Take multiple measures, and provide sufficient breaks and cold drinking-water.

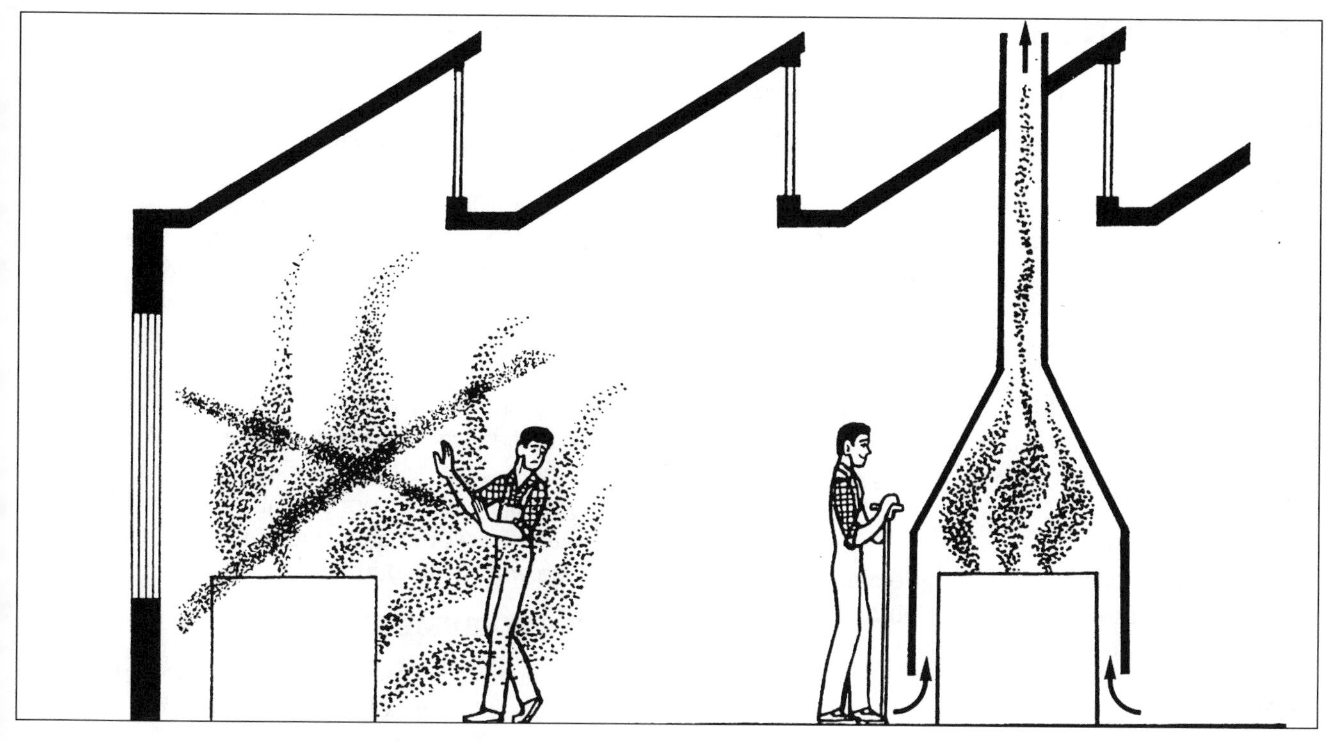

Figure 82a. Use of local exhaust against heat radiation and pollution

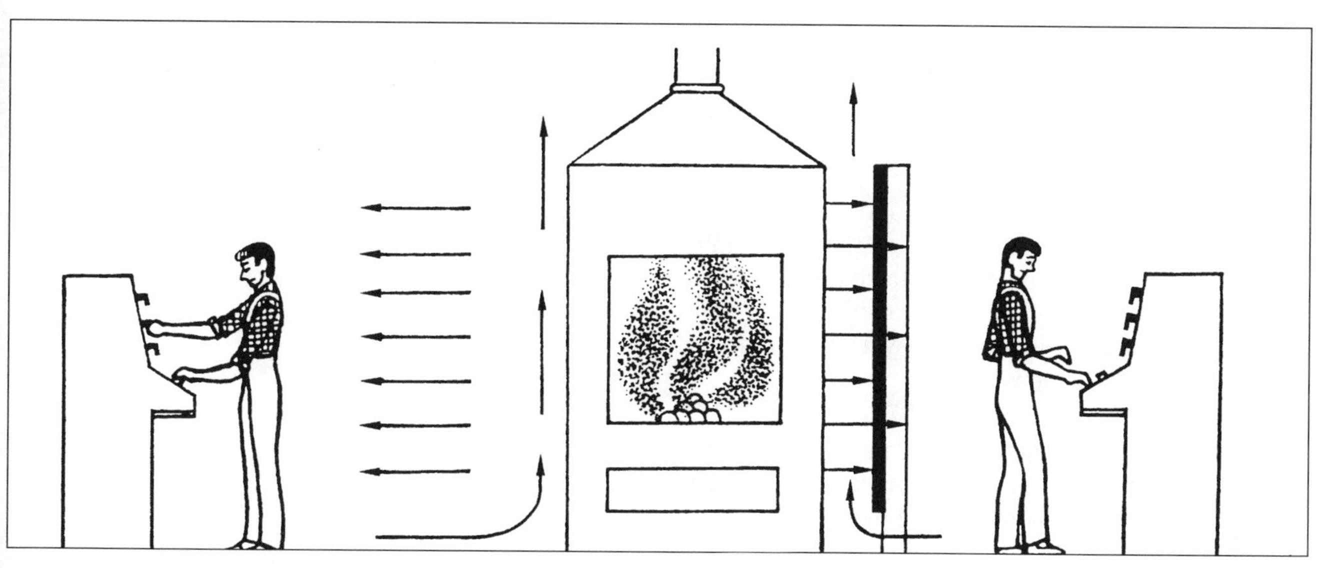

Figure 82b. Use of an absorbent shield to block heat radiation

CHECKPOINT 83

Protect the workplace from excessive outside heat and cold.

WHY

If the heat from outside is too strong, as during a hot summer or in the tropics, one of the first things to do is to reduce the amount of sunlight coming into the workplace.

If the outside is cold, as in winter, it is important to stop an unnecessary inflow of air and protect the workroom from cooling down. Heating is not the only answer.

There are multiple ways of protecting the workplace from outside heat or cold. Combining these measures leads to effective results.

HOW

1. Install insulation under roofs or on walls. The insulation can prevent heat or cold from penetrating into the workplace. Good ceilings, in addition to roofs, are also useful for the same reason. Utmost care should be taken against radiation from the sun in the case of heat and against any draught of air in the case of cold.

2. Use shades, canopies, louvres and screens so that solar radiation does not heat up workrooms or walls. Shades that prevent sunshine from falling on walls are particularly useful. For example, attach light-coloured vertical screens to the outside of the walls.

3. In hot climates, paint the outside surface of roofs and walls in light colours and make their surface smooth so as to reflect sunlight more.

4. In tropical countries, vertical screens can be used effectively in combination with horizontal shades. The horizontal shades block out the sunlight when the sun is high in the sky.

5. Plant trees, bushes, flowers and grass to protect against heat (or cold) and dust from the outside. In hot climates, open, sandy and rocky areas increase heat.

SOME MORE HINTS

- Ponds and water tanks around the enterprise help provide cool air, as wind passing over water is cooled despite a hot climate.

- To further improve protection against solar heat, the use of sun-reflecting or even coloured glass is quite effective. The simplest solution is to paint the upper part of the window glass with a water-based solution of blue dye or laundry blue.

- Heat from neighbouring hot machines or hot processes is sometimes a major problem. Move the heat sources to the outside or provide good screens, heat barriers or walls with insulation to isolate the workplaces without heat sources from those with heat sources.

POINTS TO REMEMBER

Use multiple means to protect the workplace from outside heat or cold. Combined, they can be surprisingly effective.

Figure 83a. Trees and bushes are a natural protection against heat

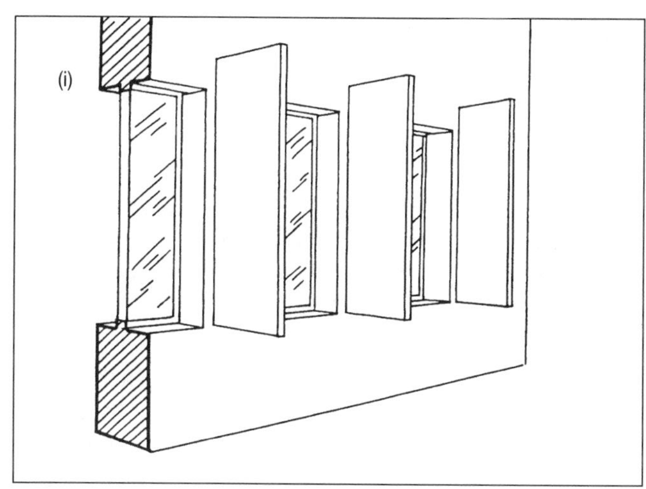

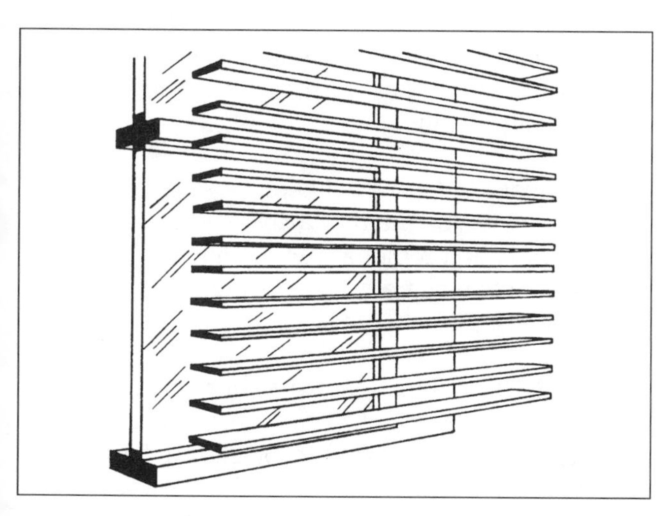

Figure 83b. External louvres made of wooden planks give all-day protection against solar radiation

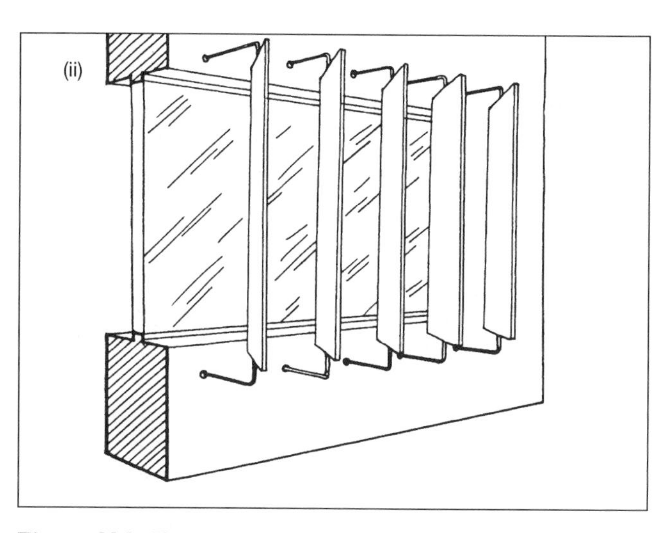

Figure 83d. External vertical screens against solar radiation. (i) Permanently fixed. (ii) Adjustable

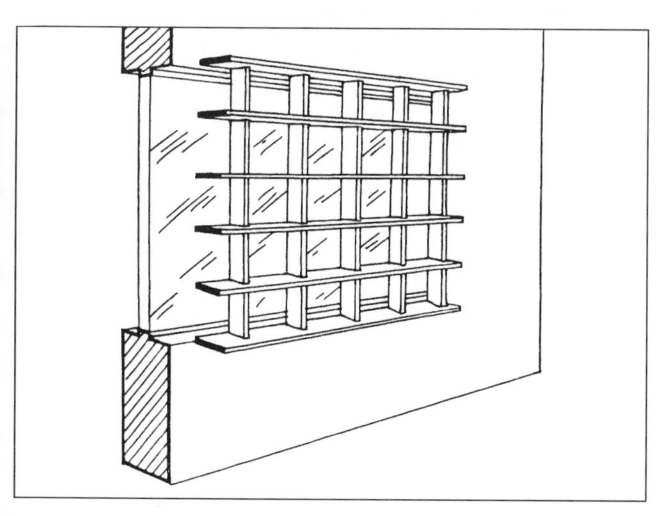

Figure 83c. External comb-type shades

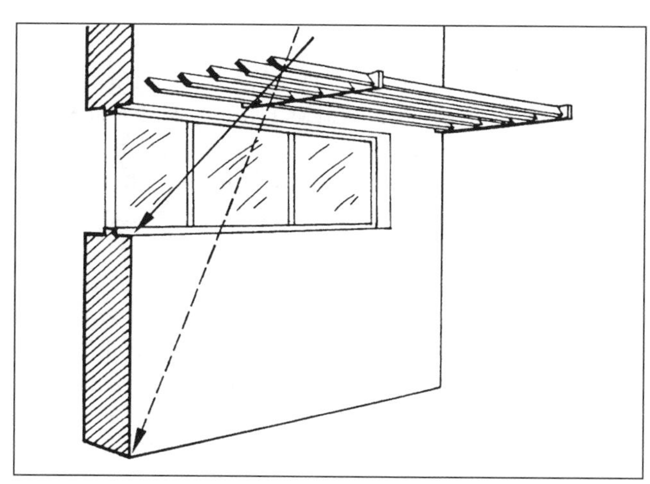

Figure 83e. A canopy made of a row of planks provides good protection against heat penetration and better lighting conditions than a solid canopy

CHECKPOINT 84

Isolate or insulate sources of heat or cold.

WHY

Hot machines or hot processes greatly add to heat stress. This is because they warm up the workroom air and because heat radiation from them directly warms up the workers.

Isolation or insulation of these hot machines or processes can reduce both the heating up of workroom air and the effects of radiation. Therefore this is a very effective way of protecting workers from heat stress.

The isolation or insulation of heat sources has a triple effect: it can keep heat in; it reduces fuel costs; and it improves work quality and comfort of workers in the workplace itself and in neighbouring workplaces.

Working in an environment exposed to cold processes also requires special protection. Cold conditions can cause frostbite, excessive heat loss and serious health consequences. Insulation can effectively prevent these effects.

HOW

1. Locate sources of heat or cold (machines and processes) outside or at least near the exterior so that heat can escape or so that cold does not have too great an effect.

2. Insulate hot or cold parts by using appropriate insulating material such as foam or polyester over metallic surfaces. Note that not all insulating materials tolerate high temperatures. However, avoid materials containing asbestos.

3. Use shields (heat barriers) between a hot oven or other radiant heat source and the worker. A good shield is made of reflective material. Water curtains or wooden or fabric-covered partitions also cut the radiating heat efficiently.

4. In the case of cold processes, insulation is the most efficient way to protect workers from injuries and heat loss.

SOME MORE HINTS

- An alternative to insulation is the use of personal protective clothing to reduce heat radiation. In tropical conditions, however, personal protection from heat is less efficient and the main emphasis is on insulation, guards and increased air flow.

- Heat-protective aprons or clothing (e.g. aluminium-coated garments) are effective against radiant heat. If, however, air temperature and humidity are high, such clothing may have adverse effects by making the worker very uncomfortable, because perspiration is not allowed to evaporate. Working in hot conditions can become dangerous if measures are not taken to prevent heat exhaustion. In this case, additional measures may be necessary to provide a strong or constant air flow inside the protective clothing.

- Automation of tasks in hot environments may in some cases be the only solution to avoiding heat-related problems. Air-conditioned booths can sometimes be one of the best available solutions.

- In cold conditions, the use of vibrating tools increases the risk of white-finger disease. Hands and feet need particular protection. If through insulation or good working clothes the whole body is warm and comfortable, hands and feet also remain warm.

POINTS TO REMEMBER

Insulate the surfaces of hot machines and processes, and use heat barriers to prevent radiant heat from reaching the worker. Both of these greatly reduce heat stress in workers.

Figure 84a. Where exposure to excessive heat sources is unavoidable, minimize the exposure time and wear clothing that can protect workers from heat radiation. Remember to take measures to prevent heat exhaustion

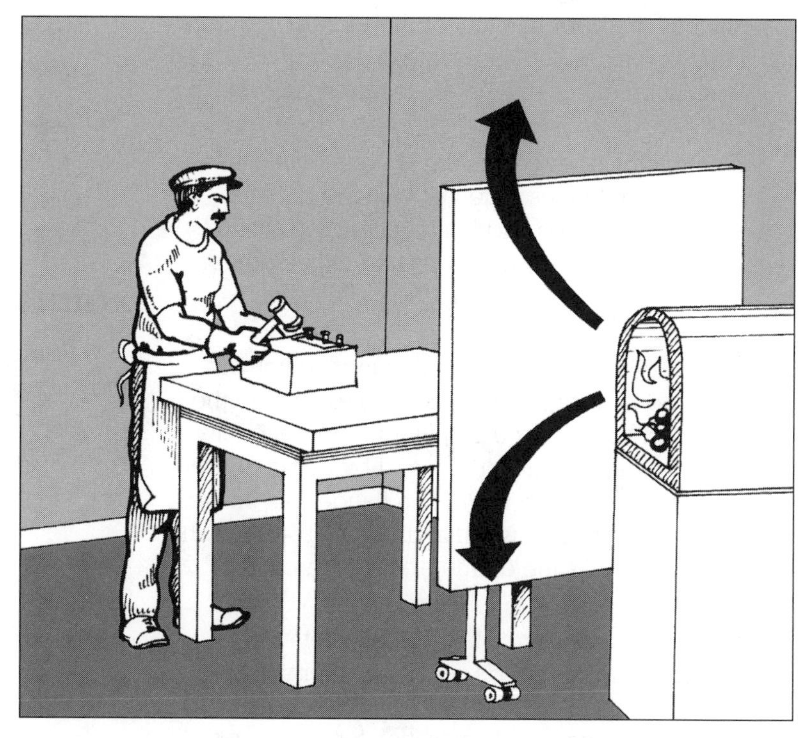

Figure 84b. Heat barriers prevent radiant heat from reaching the worker, which reduces heat stress

CHECKPOINT 85

Install effective local exhaust systems which allow efficient and safe work.

WHY

Hazardous substances in the air are a source of ill-health. If workers fall sick as a result of being exposed to such substances, it will cost you much the same as accidents do. Even before the workers suffer from diseases, exposure to hazardous substances may cause fatigue, headache, dizziness and irritation of the eyes and the throat, and workers cannot work efficiently. Absenteeism and turnover of workers may increase. Local exhaust is an effective way to prevent all these problems.

When exhaust ventilation is used, it is important to use suitable types of hood or flange at appropriate places. Otherwise, polluted air may have difficulty in being expelled from the workplace.

HOW

1. Use an enclosing-type exhaust system if the substances are very harmful or the area to be ventilated is narrow. You can achieve the highest ventilation effect with an enclosing-type system.

2. If an enclosing-type system is not realistic, use hoods and flanges together with the exhaust ventilators. Hoods and flanges limit the air flow from unnecessary directions and thus increase the efficiency of collecting the polluted air.

3. Use correct types of hood or flange which are installed at appropriate places in relation to the pollution source. By using flanges, the capacity of exhaust ventilation is increased by about 25 per cent. The adequate width of a flange to achieve efficient ventilation is a maximum of 15 cm or the same length as the diameter of the duct.

4. Provide shutters or curtains to the inlets of exhaust ducts or hoods, and close them when the ducts or hoods are not in use. You can increase the power of ventilation for other work areas where this is needed. You can also increase the efficiency of ventilation by narrowing unnecessary parts of the inlet with a shutter or curtain.

5. Use shields, partitions and barriers to increase ventilation efficiency from the pollution source to the inlets of the exhaust ducts. Close windows near the exhaust inlet or provide partitions around the hoods to avoid the disturbing effects of draughts.

6. Use a portable suction exhaust if the pollution source changes as the worker performs the work that produces hazardous substances (e.g. welding).

SOME MORE HINTS

- Different types of cover or hood with built-in extractors can be connected to hand-operated machines such as grinders. Place a cover or hood so that flying particles can automatically and easily be sucked into the extractors.

- Place hoods considering air flow. Hot air rises, so hoods above the heat sources are more efficient than others.

- If the hood is set in the direction of natural air flow or the expected flow of hazardous gases, the efficiency becomes high. Select the proper place for exhaust hoods.

- If a local exhaust system cannot be arranged or the existing exhaust system is still insufficient, the workers should be provided with adequate respirators.

POINTS TO REMEMBER

Removing dust and gases at the source before they reach the workers is usually the most cost-effective way to prevent their inhalation. Use effective local exhaust systems if enclosing systems cannot be applied.

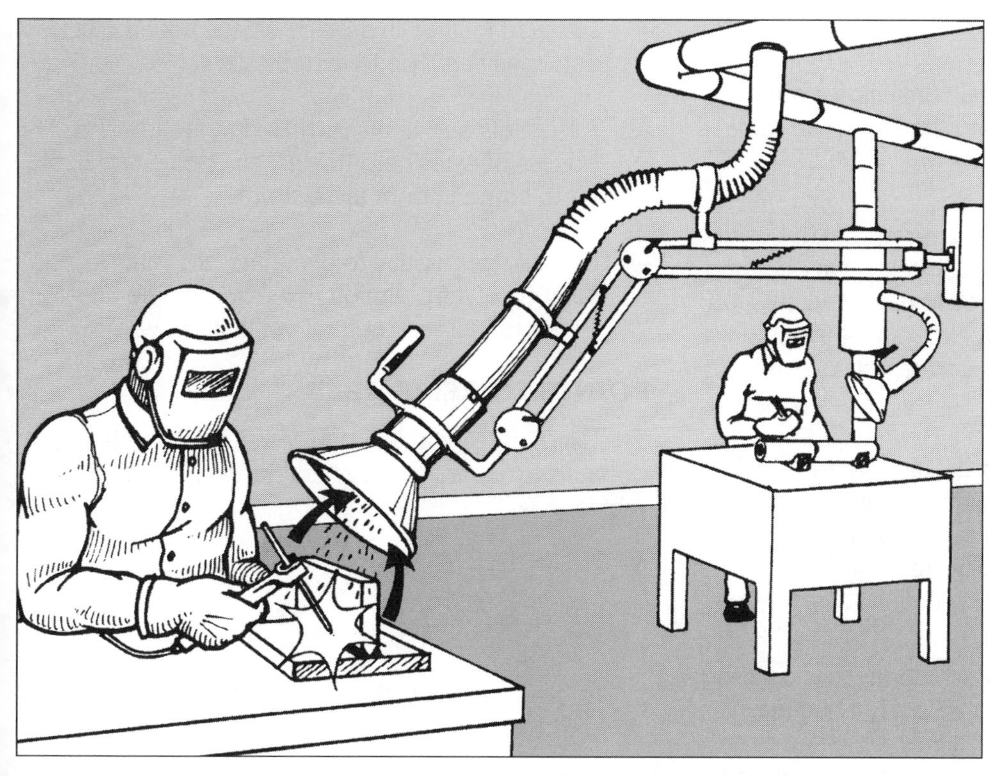

Figure 85a. When enclosing systems cannot be applied, use local exhaust systems to remove dust and gases at source before they reach the worker

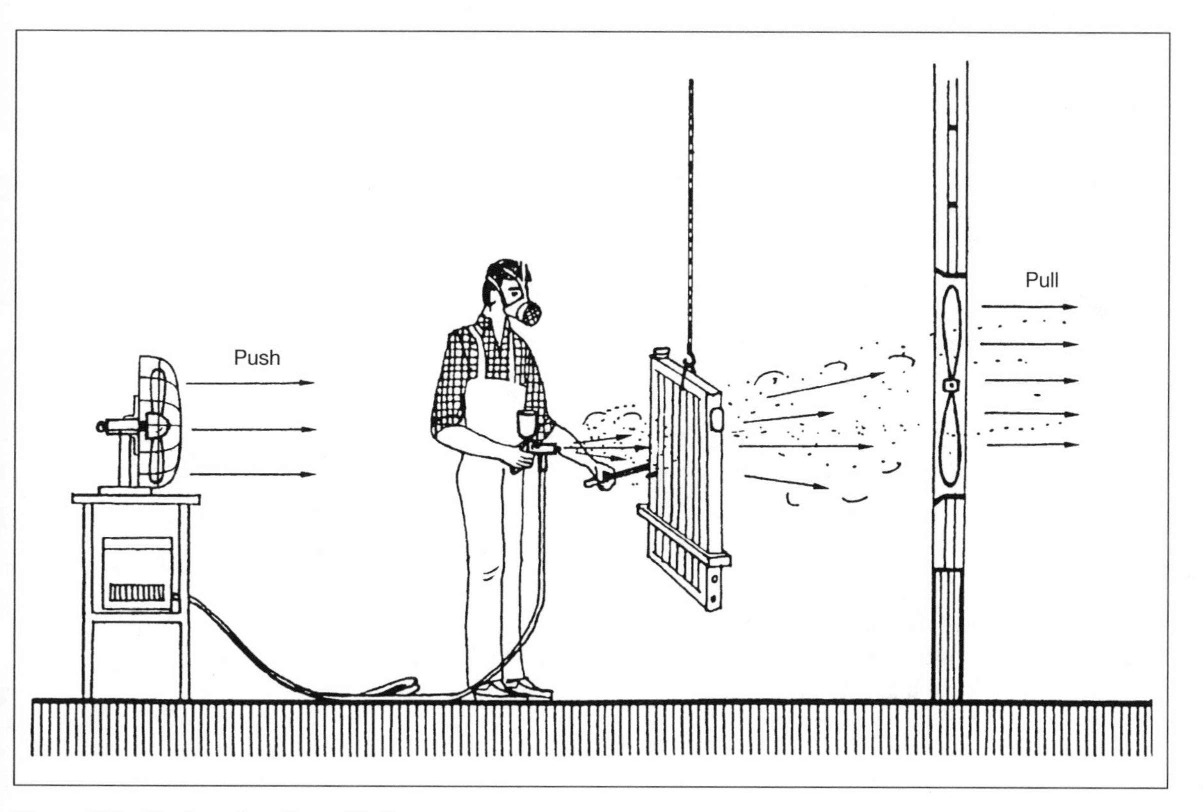

Figure 85b. Push-and-pull ventilation

CHECKPOINT 86

Increase the use of natural ventilation when needed to improve the indoor climate.

WHY

Natural air flow serves as a very powerful ventilator. Winds bring in fresh air and take away hot and polluted air. Measures to increase natural ventilation are generally recommended, except in very cold weather or in processes unsuited to outside air or to changing air flow.

Before installing costly ventilation systems, an increased use of natural ventilation is an option worthy of consideration. Natural ventilation can be combined with the use of ventilation equipment.

HOW

1. Choose a place for work where there is a strong natural air flow, in particular in a hot climate. Avoid doing work in a relatively enclosed corner or in narrow places surrounded by equipment or partitions.

2. Increase openings facing the outside, for example, by opening windows and doors or creating new openings. All these help increase natural ventilation.

3. Rearrange equipment if it disturbs the natural air flow and relocate or remove partitions.

4. Provide or relocate windows and openings in higher positions or use ceiling fans in order to increase the hot air flow directed towards the outside (hot air rises).

5. Establish a practice of opening windows partially or as a whole depending on the weather and winds.

SOME MORE HINTS

- When relying on increased natural ventilation (e.g. in hot climates), it is important to protect the workplace from outside heat. It is equally important to move heat sources to the outside of the workplace and to improve production procedures so as to minimize the need for special ventilation.

- Install machines in a place where hot air can rise and easily escape to the outside.

- Air outlets and inlets provided at the top and bottom of walls help heated air to rise and cool air to come in from the bottom.

- It is also necessary to eliminate or isolate sources of air pollution at the same time as trying to increase natural ventilation.

POINTS TO REMEMBER

Use natural ventilation fully, especially in hot climates, as it carries away hot air and polluting substances.

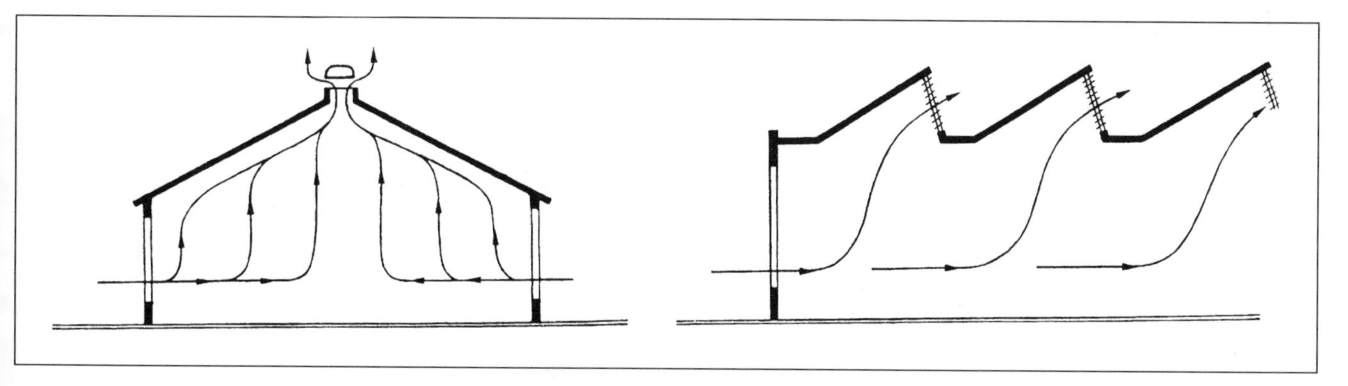

Figure 86a. Air-flow routes in buildings of different design

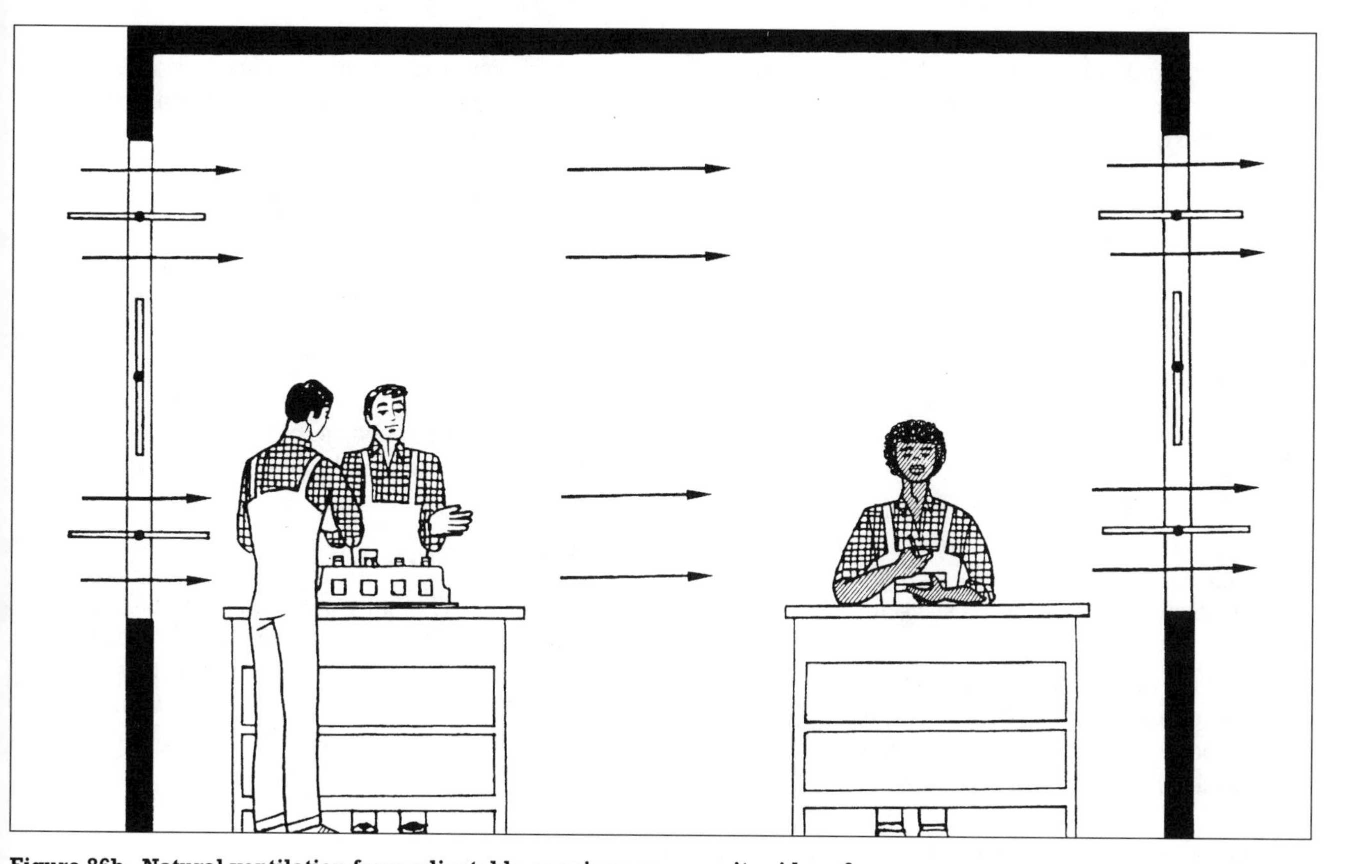

Figure 86b. Natural ventilation from adjustable openings on opposite sides of a room

CHECKPOINT 87

Improve and maintain ventilation systems to ensure good workplace air quality.

WHY

Good air flow in a workplace is very important for productive and healthy work. Adequate ventilation can help control hazardous substances and prevent accumulation of heat.

The efficiency of a ventilation system depends not only on its overall capacity but also on how air flow is created and how polluted or heated air is collected. Simple changes in the location of ventilators, electric fans and hoods, or rearrangement of the location of work areas, can often bring about remarkable benefits.

Polluted air is a problem when it is inhaled. Air flow coming from the worker to the polluting sources (and to the outside) greatly helps to reduce this problem. The direction of ventilation must carefully be taken into account.

HOW

1. Choose a ventilation system that can effectively ventilate the whole work area for which it is installed. Consult a supplier with good knowledge and experience of such a system.

2. Position ventilation duct inlets and outlets or fans where they are most efficient for the ventilation purpose. Consult ventilation manufacturers or specialists.

3. Use both push-type and pull-type ventilation in a good combination. Place push ventilation in areas where there is no danger of polluting other places and use pull ventilation at or near polluted worksites. The capacity of an exhaust pull fan must be 5-15 times larger than that of a push-type fan.

4. Hot air rises, so the use of ceiling fans and windows in higher positions can improve ventilation efficiently. Remember that a chimney has no power source but can ventilate smoke efficiently.

5. Establish a practice of opening windows, as it is a popular and simple way to increase cross-ventilation. Multi-section windows can help to control the air flow according to the wind strength.

SOME MORE HINTS

- When working near pollution or heat sources, the position of the worker should be such that air blows from the worker to the pollution or heat sources (and not in the opposite direction). Avoid air-scattering push-type ventilation towards such sources.

- If installing a good ventilation system for the whole workplace is not realistic (e.g. because of hot machines or strong dust sources), isolate part of the work area by means of partitions and inner roofs, and provide air-conditioning for it.

- Do not expect ventilation systems alone to prevent pollution. Take measures to eliminate or isolate the sources of pollution (e.g. by removing them from the workshop to a place outside under a canopy).

- Maintain ventilation systems by putting individuals or firms in charge of them, and regularly clean floors, walls and machine surfaces appropriately.

POINTS TO REMEMBER

To achieve efficient ventilation, use push-type and pull-type ventilation properly. When you work with hazardous chemicals or heated process, an air flow that goes from the worker to the source of pollution or heat is important. Use both common sense and specialist assistance.

Figure 87. A combined ventilation system. (i) Exhaust fan. (ii) Louvred skylights

Control of hazardous substances and agents

CHECKPOINT 88

Isolate or cover noisy machines or parts of machines.

WHY

At many workplaces, the level of noise created by machines can be injurious to hearing and can affect the health of workers (a noise level of 85-90 dB(A) or more is harmful to hearing). If you stand at arm's length from your co-worker and cannot communicate in a normal tone of voice, the noise level is too high.

Levels of noise that are too high can cause accidents and affect production, as warning and other signals are not heard.

The best way to reduce noise is to enclose entire machines or particularly noisy parts of machines.

If the noise cannot be reduced at the source, you can still consider isolating the noisy machines away from the places where work is actually done.

HOW

1. Enclose entire machines that produce excessive levels of noise. If this is not possible, enclose particularly noisy parts of machines.

2. If possible, position particularly noisy machines outside the workplace and cover them with appropriate structures.

3. Relocate particularly noisy machines so that they are at a distance from the place where most of the workers are working. While the workers operating the noisy machines require ear-muffs, other workers may no longer be exposed to harmful noise.

4. Provide screens or partitions to isolate noisy machines from other workers in the same workplace. For a noisy machine, providing a booth that can cover the whole operating site, or ceiling-high partitions, can be quite effective.

SOME MORE HINTS

- Most machines have particularly noisy moving parts. List such sources of noise and discuss with workers if these noisy parts can be covered without disturbing their operations.

- Covers of noisy machines must be tight enough. Make sure that these tight covers do not cause overheating inside the cover.

- Screw-on mufflers are effective in reducing the noise from pneumatic exhausts.

- Change the type of machine that is particularly noisy. There are many new types which are much less noisy.

POINTS TO REMEMBER

If you are unable to speak in a normal tone of voice standing at arm's length from your co-worker, then the noise level is harmful to hearing. Steps must be taken to keep your ears screened from the noise, either by enclosing the noise source or by wearing ear-protectors.

Figure 88a. Protect your ears by enclosing, isolating or covering noisy machines, or by wearing ear-protectors

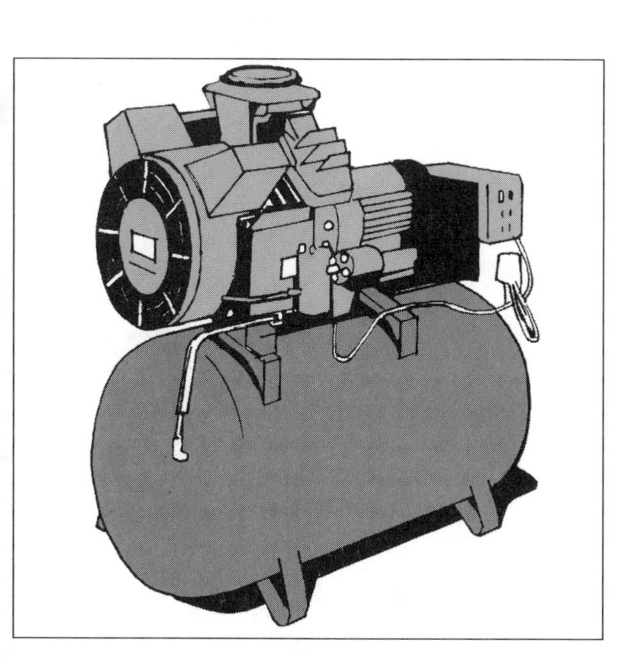

Figure 88b. Noise-insulated air compressors. The principle is that the noise should be contained under the hood. The hood is made of hard material with a soft, absorbent lining

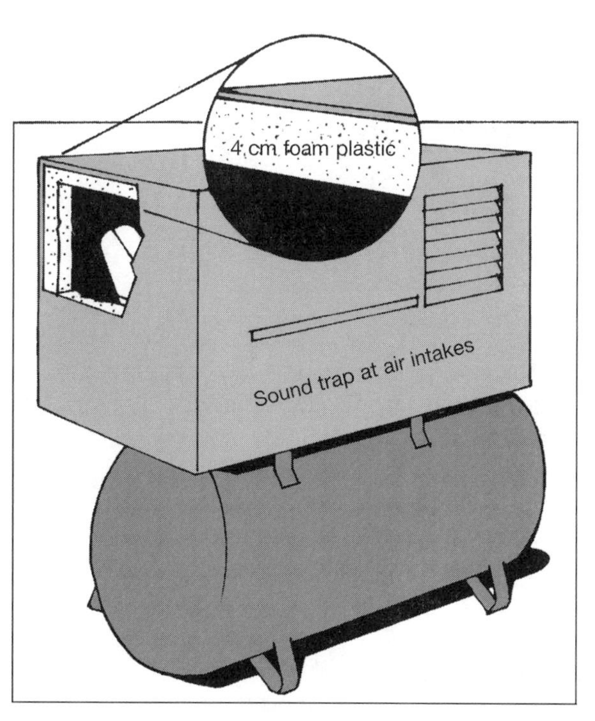

Figure 88c. 1.5 mm stiffened plate reduces vibrations

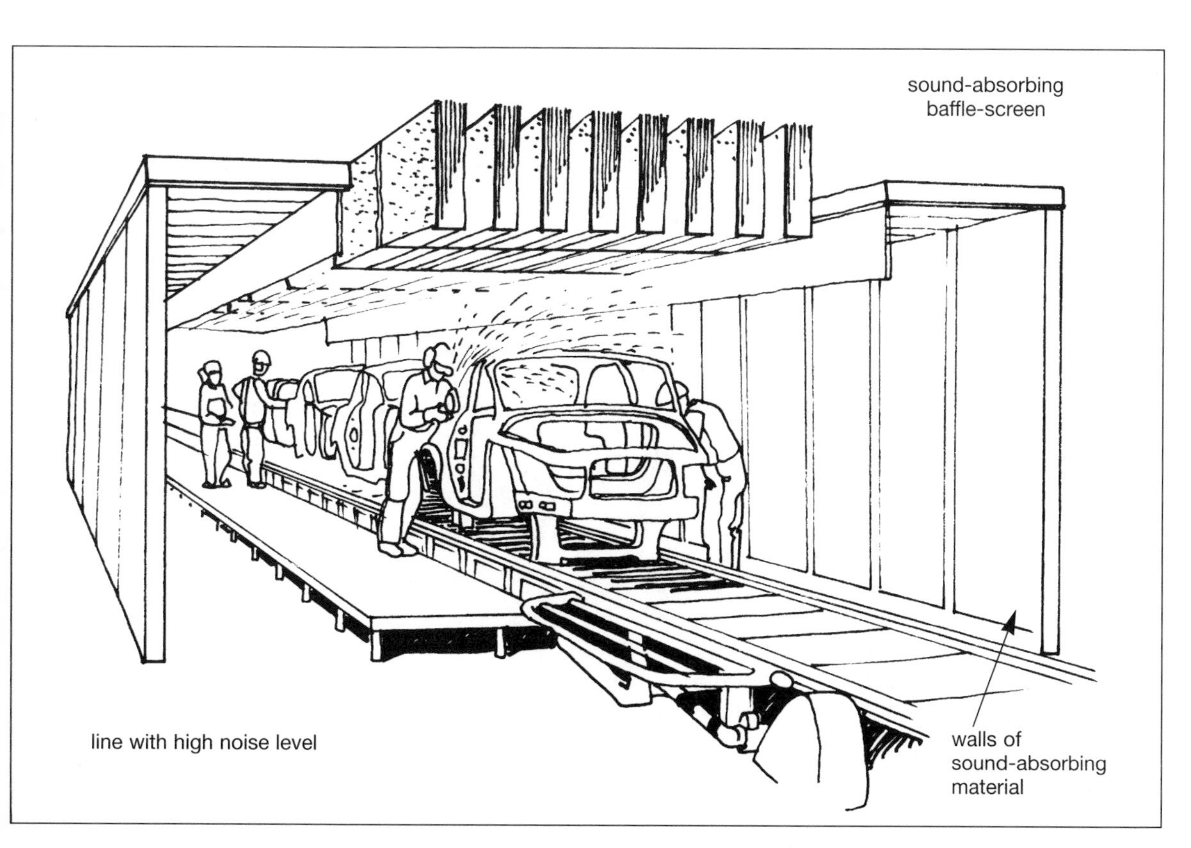

Figure 88d. For noisy production lines, walls of sound-absorbing material and the use of sound-absorbing baffle-screens are useful to reduce the noise level and separate the production lines from neighbouring areas in order to prevent the disturbing effects of noise from affecting those areas

CHECKPOINT 89

Maintain tools and machines regularly in order to reduce noise.

WHY

Often noise levels from tools and machines increase because of poor maintenance or unnecessary vibration. Regular maintenance can greatly help reduce noise levels.

Noise can come from loosely fastened parts or metal parts struck by materials. Such noise can be easily reduced by proper maintenance.

HOW

1. Establish a rule for servicing tools and machines regularly to keep them in good condition and thereby reduce the noise.

2. Check if vibrations of certain components of the machine or metal casings are causing unnecessary noise. Maintain these parts properly. For example, make sure that noise-producing parts or casings are properly fastened.

3. Replace metal parts by parts made of sound-absorbing material, e.g. plastic, rubber or other soundproof materials.

4. Cover the ceilings and walls with sound-absorbing materials. Also check whether sound-absorbing screens are properly placed.

SOME MORE HINTS

- If appropriate, reduce sharp blows by lengthening the braking period for reciprocating parts or by using rubber or plastic coverings.

- Reduce the unnecessarily high speed of noise-producing power transmission parts or conveyor systems.

- Noise generated when materials hit a chute can be easily reduced by covering the outer surface of the chute with soundproof materials.

POINTS TO REMEMBER

Good maintenance can reduce the amount of noise from tools and machines. Experienced workers can tell you how to keep them in good condition.

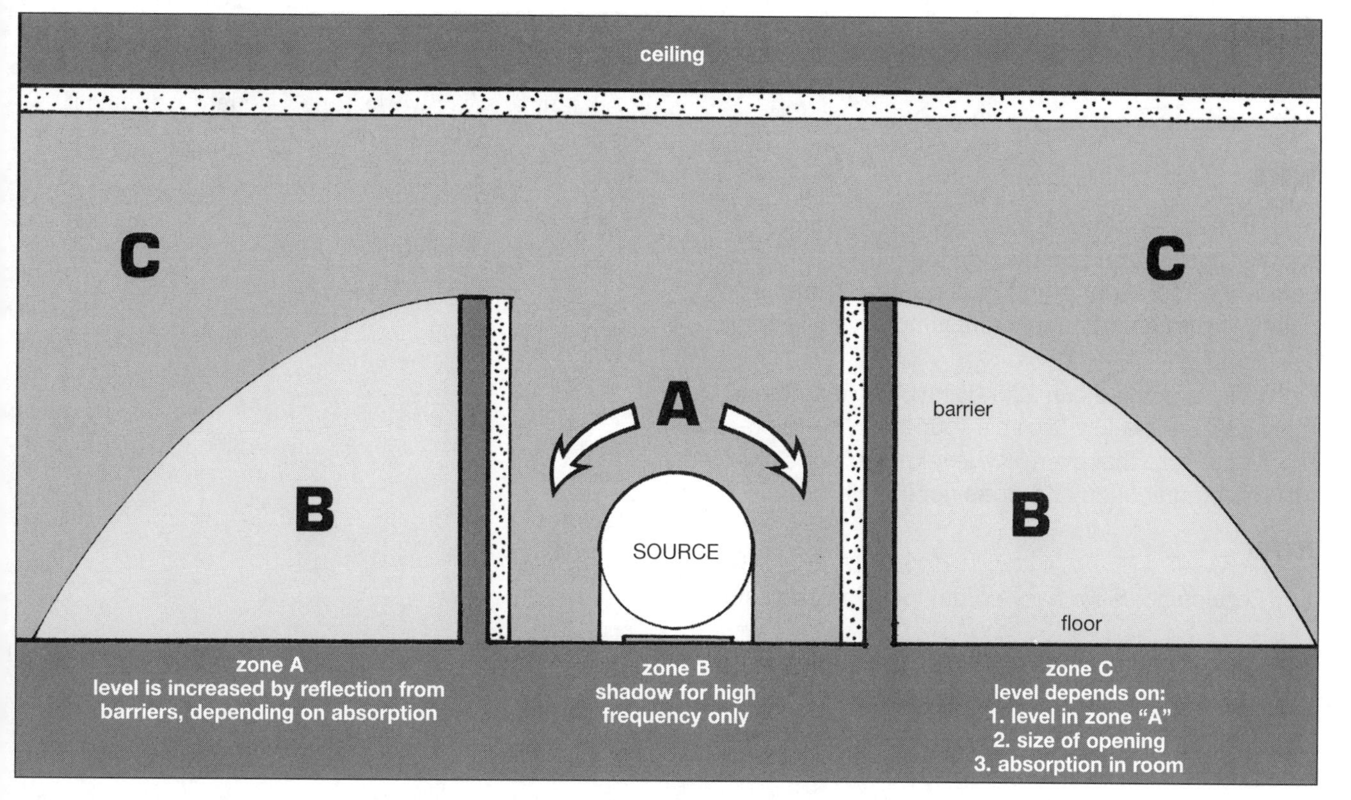

Figure 89. Check if sound-absorbing barriers are properly placed to reduce the noise level at the work area behind the barriers

CHECKPOINT 90

Make sure that noise does not interfere with communication, safety and work efficiency.

WHY

High levels of noise interfere with warning shouts, signals and communication. This can cause accidents and affect production quality. Communication is particularly important in noisy workshops.

Irritating noise can also disturb work and lead to mistakes. Even a low-volume sound can be distracting. Make sure that even low-level noise does not disturb communication and does not irritate people.

HOW

1. Reduce noise for easy communication and safety.

2. Check with workers whether communication essential for work and necessary warning is impeded by noise. Consider alternative means of communication.

3. Use higher levels of warning signals or add lamp signals.

4. Provide partitions or noise-proof booths for worksites where communication with others plays an important role.

5. Provide partitions around telecommunications equipment or use better equipment to ensure good communication within the existing noise levels.

SOME MORE HINTS

- Install soundproof material in the ceiling and walls near the worksites requiring frequent communication.

- Place light signals near eye height so that they can be seen easily when lit. For warning signals, however, use acoustic signals that are loud enough, because lights are only seen when the workers look in that direction.

- Make sure that necessary communication reaches the workers' ears when they are using noise-protectors.

POINTS TO REMEMBER

Proper means of communication are particularly important in noisy workshops. Use lamp signals if needed.

Figure 90a. Make sure that necessary communication is not disturbed by noise

Figure 90b. Make sure that noise protectors are effective but do not shut out necessary communication. Emergency light signals and alarms are important means of communication when hearing protection is being used

CHECKPOINT 91

Reduce vibration affecting workers in order to improve safety, health and work efficiency.

WHY

Many machines or power-driven hand tools transmit their vibrations to the human body. Together with the noise, these vibrations can be harmful. They can injure muscles and joints, and affect blood circulation. "White finger disease" seen among workers using pneumatic drills or chain-saws is a painful example.

Whole-body vibrations of workers on vehicles, cranes or forestry machines are also a problem. They can cause considerable discomfort, difficulty in seeing objects accurately and even damage to internal organs.

These vibrations are usually difficult to control. But machines are being developed which vibrate far less than before, and there are various inexpensive means of dampening vibrations.

HOW

1. To reduce hand-arm vibrations, purchase equipment with vibration-isolated handles. Study tool specifications before ordering the equipment.

2. To reduce vibrations of existing tools, cover handles with vibration-insulation foam and provide vibration-absorbing gloves. Check with workers if vibrations have really diminished.

3. In using vibrating tools, avoid continuous vibration and rest the tool on a support or workpiece as much as possible.

4. Reduce the vibration of tools, machines and vehicles by improving maintenance.

5. To reduce whole-body vibrations, isolate the body by better seat suspension, cushions on the seat, etc.

6. Rotate people within a day to reduce exposure to vibration per worker.

SOME MORE HINTS

- Combine the work exposed to vibration with non-vibrating tasks, or insert short breaks.

- Minimize hand-grip force on vibrating tools so far as it is consistent with safe working.

- In a cold climate, keep the body and hands warm and dry while using vibrating tools.

POINTS TO REMEMBER

Reduce vibration effects by better engineering and better management. Also improve equipment maintenance.

Figure 91. Reduce the vibration of tools and machines by improving their maintenance

CHECKPOINT 92

Choose electric hand lamps that are well insulated against electric shock and heat.

WHY

Although hand lamps are useful for work done while moving from one place to another, they are among the most dangerous portable appliances in any workplace. They are often used in wet and humid places where the risk of electric shock is increased.

Portable electric lamps are mainly used in moving (ambulatory) tasks and in confined locations. The fixtures and cables tend to wear out rapidly, thus increasing the risk of electric shocks.

Unshielded electric lamps may cause burns and may indirectly lead to stumbling or falling from ladders or scaffolding.

HOW

1. Select electric hand lamps which are well insulated and have steady fixtures and cables made of isolating material, and which are resistant to heat and chafe.

2. Verify before each use that the unit is grounded unless it is safeguarded with a transformer for low-voltage use.

3. Verify that the protective frame around the hand lamp is in place and secured, and that the metal grid does not make contact with the metal parts of the lamp holder.

4. Arrange regular checking of portable electric devices, including hand lamps. Make known to all workers who is responsible for inspection and maintenance.

SOME MORE HINTS

- Use a fixture that has a good protective grid, one side of which is covered with a reflector-shield to prevent glare.

- A portable fluorescent lamp is safer than other lamps in terms of preventing electric shock. It gives more light that is more evenly distributed.

- Verify that the screw-cap lamp holder is protected so that it is impossible to touch the screw-cap while it is still in contact with the female thread on the holder.

- Provide a home for each hand lamp to ensure organized storage and maintenance.

POINTS TO REMEMBER

Maintain electric hand lamps to prevent electric shock and burns.

Figure 92. Use hand lamps that are well insulated and have steady fixtures and safe cables

CHECKPOINT 93

Ensure safe wiring connections for equipment and lights.

WHY

Wiring and light connections are a major source of accidents due to electricity, in particular electric shocks. Special care must be taken to prevent irregular wiring and damages to connections.

Good maintenance of wiring and connections can minimize loss of time and interruptions due to equipment failure. Good maintenance can also reduce electrical accidents.

HOW

1. Insulate or guard electrical terminals. Ensure that all wirings are appropriate.

2. Provide a sufficient number of socket outlets for wiring connections in order to minimize contacts of workers with cables. If necessary, use an additional multi-plug socket block.

3. Use only prescribed connections and eliminate unauthorized wiring. Never use exposed connections. Inform and train workers about the use of appropriate wire gauges which match the electric power required for machines, equipment and light units.

4. Provide proper grounding for machines and equipment, and ensure that power tools and hand lamps in use are grounded.

5. Make it a rule to replace frayed cables quickly. Ensure strict observance of this rule by users of power tools and hand lamps.

6. Train all workers in how to work safely with electric circuits and connections.

SOME MORE HINTS

- Protect electric circuits and cables from accidental leakage or spillage of liquids.

- Establish for each workplace a programme of regular inspection of electric circuits and portable electrical equipment.

- Protect cables, especially those temporarily placed on the floor, from being stepped upon by workers or wheeled over by transport equipment. Place a stable protective cover where there is such a danger.

- Remember that laws and regulations include requirements concerning electric installations. Carefully study them with workers with a view to following them precisely.

POINTS TO REMEMBER

Safe wiring and electrical connections result in less chance of fires and time lost due to machine failures or injury of workers.

Figure 93a. Safe wiring and electrical connections prevent accidents and fires

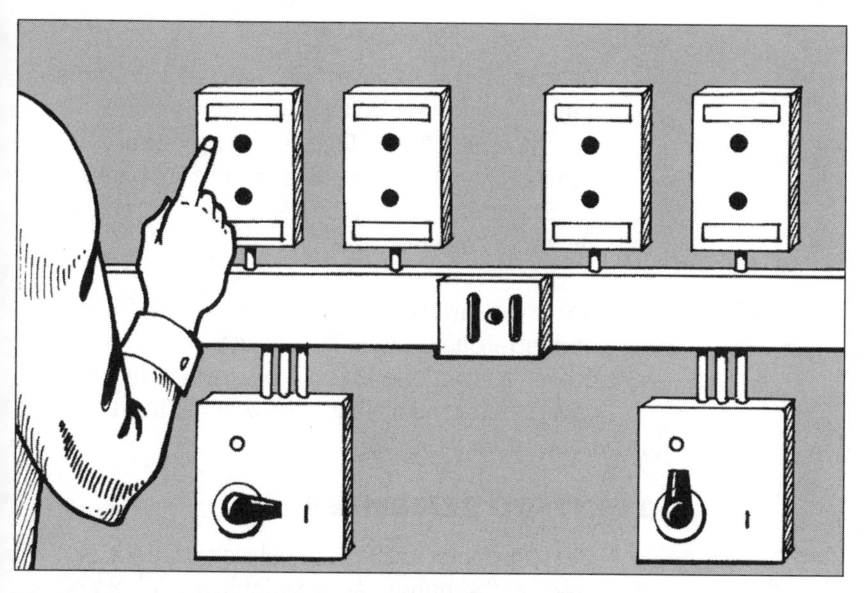

Figure 93b. Keeping wiring and electrical connections well maintained can minimize lost work time due to equipment failure

CHECKPOINT 94

Protect workers from chemical risks so that they can perform their work safely and efficiently.

WHY

Chemicals can seriously injure workers. Common examples include paints, solvents, cleaners, acids, pesticides and gases. To prevent injury, you must have precise information about the risks and necessary countermeasures.

Exposure to chemicals affects workers' performance and accuracy. Incorrect disposal of chemicals can harm the environment outside the workplace. Necessary precautions must be taken from the entry of the chemicals to the enterprise until their disposal.

Many chemicals have long-term effects that are not readily seen. The compensation costs might be very high. Prevention is much less costly.

HOW

1. Select equipment and processes to minimize chemical risks to workers. Wherever possible use less hazardous chemicals.

2. Label containers of all hazardous chemicals. When the chemicals are transferred to other containers, do not forget to label the new containers.

3. Provide each person working with hazardous chemicals with written instructions containing illustrations. Safety instructions and chemical safety data sheets should be in languages easily understood by the workers and readily accessible at the workplace.

4. Provide training to workers using chemicals. Do not rely only on written materials in the training.

5. If possible, enclose the source of hazardous chemicals so that the workers are not exposed to the chemicals. Or locate the worksite as far away from the sources as possible.

6. If enclosing the entire process is not possible, use covers, hoods or booths connected to exhaust systems.

7. If needed, provide workers with sufficient personal protective equipment (such as protective clothing, goggles, gloves, respirators and work boots).

SOME MORE HINTS

- Inform the workers and management about the latest laws and regulations regarding the use of chemicals in the workplace.

- Check equipment and processes for leaks.

- Specifications in the manufacture of products such as paints, varnishes and adhesives are now calling more and more for a water base rather than a solvent base.

- Total enclosure or local extraction of polluted air needs to be supplemented by increased general ventilation.

- Different types of cover with built-in extractors can be connected to tools and some hand-operated machines. For welding, there are a number of smaller ventilation units that either can be connected to the central air extraction system or are portable.

- An air-curtain system can be arranged for baths of dangerous liquids. The air is blown in under pressure from one side and extracted from the other, thereby screening the worker from the dangerous vapours.

POINTS TO REMEMBER

Certain dangerous chemical risks cannot be detected by the human senses. Inform workers of these risks and train them about necessary precautions. Protecting workers from hazardous chemicals is much less costly than compensation costs resulting from exposures.

Figure 94a. Noxious fumes can be drawn into a local exhaust system to prevent the worker from being exposed to chemical hazards

Figure 94b. Drawing contaminants into the work table before they can reach the breathing zone of the worker is another method of local ventilation

Welfare facilities

CHECKPOINT 95

Provide and maintain good changing, washing and sanitary facilities to ensure good hygiene and tidiness.

WHY

Well-maintained washing facilities, toilets and lockers meet some of workers' most essential needs. These basic facilities, sufficient in number and maintained at good levels of hygiene, represent the "face" of your enterprise.

Washing facilities that are conveniently located help to prevent chemicals from being absorbed through the skin or being ingested during snacks and meals. Dirt and grime are unpleasant and may also cause sickness.

Good washing facilities and clean toilets further help to maintain a good working atmosphere.

Facilities for secure storage of clothing and other personal belongings also greatly help to maintain personal hygiene.

HOW

1. Check whether washing facilities, toilets, and lockers or changing-rooms are far from the worksite, insufficient in number or poorly maintained.

2. Make plans for improving existing facilities with respect to their numbers, convenience of location and design. Keep in mind that many improvements can be made at relatively low cost.

3. The legal requirements give you a guide, but providing more facilities might perhaps be useful (e.g. although the requirements may be different from country to country, practical minimum levels are: one toilet for up to five men and two for six to 40 men; one separate toilet for up to five women and two for six to 30 women; one wash-basin for every 15 workers; changing-room with lockers and showers if work is hot and dirty, requires uniforms or protective clothing, or involves chemicals).

4. When workplaces are rearranged or newly built, include good sanitary facilities and changing-rooms in the plan. This often proves cheaper in the end.

5. Establish practical arrangements to clean and maintain all these facilities.

SOME MORE HINTS

- Neglect is the main source of problems with sanitary facilities in many workplaces. Give them priority.

- The design of sanitary facilities makes a large difference to the cost and effort required for cleaning. Use floors and walls made of durable materials that are easy to clean (e.g. tiles). Special care should be taken as to proper drainage.

- Lockers should be arranged in such a way that clothes and personal belongings can be kept safe from damage and theft. They should be placed within a special cloakroom area or a changing-room, and should preferably be located as far as possible from workstations.

POINTS TO REMEMBER

Essential facilities such as washing facilities, toilets and changing areas or rooms are quite often neglected. Make sure that these facilities serve their purpose and are kept clean. They represent the "face" of your enterprise.

Figure 95a. Personal hygiene is very important in terms of reducing health hazards when using chemical products such as epoxy, isocyanides, lead and pesticides. Do not allow your dirty clothes to spread hazardous substances to your own home and family

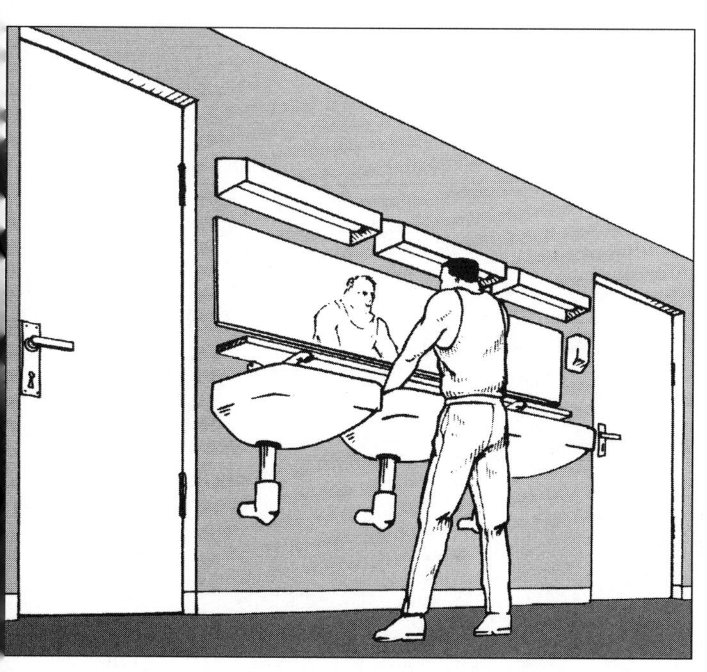

Figure 95b. Providing and maintaining good washing and sanitary facilities is essential to good hygiene and tidiness in the workplace

CHECKPOINT 96

Provide drinking facilities, eating areas and rest rooms to ensure good performance and well-being.

WHY

Good drinking facilities, eating space and rest rooms can do much to prevent fatigue and maintain workers' health.

Workers spend a substantial part of their everyday life at the enterprise. Just as they do at home, they also drink, eat and rest at the workplace. Do not forget that drinking facilities, eating areas and rest rooms are an essential part of your enterprise.

Especially in a hot environment, work results in considerable loss of water. Providing clean drinking-water is essential for all types of work.

HOW

1. Place water containers near each group of workers, or provide water taps or drinking fountains in a place with easy access (but not near dangerous machines, not in places where water can be contaminated by dust or chemicals, and not in washrooms or toilets).

2. Provide an eating area or room in which workers can eat food in a comfortable, relaxing atmosphere (away from their workstations).

3. Provide rest areas away from workstations and free from disturbances such as noise, dust and chemicals. As a minimum, a table and chairs or sofas are needed. Avoid bright sunlight.

4. Maintain hygienic conditions in all these facilities. In an eating space or a rest area, easy access to clean water for washing, drinking-water or other beverages, and rubbish bins is also important.

SOME MORE HINTS

- It is also important to arrange for the drinking-water to be cool. If a water-cooling device is not available, place the water in the coolest place.

- It is advisable to set up the eating area or room in such a way that it can be upgraded to contain some cooking facilities or into a small buffet or canteen as resources become available (e.g. a lunch-room can include a small area where workers can prepare drinks or heat their food).

- There are many different inexpensive ways of providing drinking-water, eating space and rest areas. Solutions suited to your workplace could be sought by getting feedback from workers.

- The space needed for setting up an eating area is often less than you might expect; for 50 workers 25 square metres are sufficient if workers share the space by eating in different sittings.

POINTS TO REMEMBER

Choose types of arrangement suited to your workplace for drinking, eating and resting. This greatly helps to reduce fatigue and maintain productivity and health. Clean facilities are appreciated by all workers.

Figure 96a. Drinking facilities and eating areas are important for maintaining the health of workers and are an essential part of the enterprise

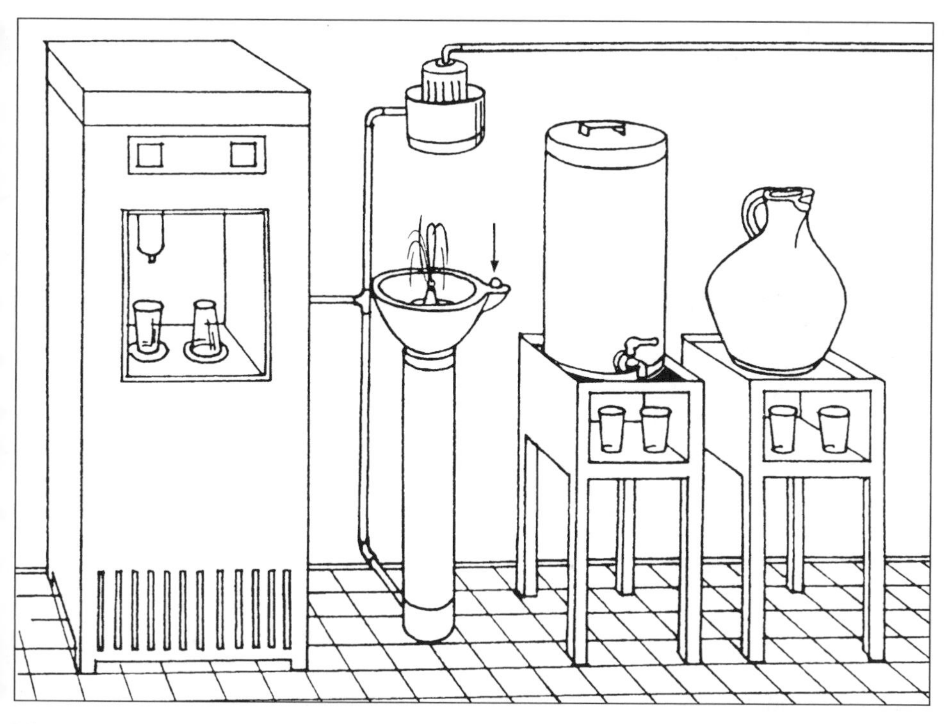

Figure 96b. Ways of providing cool, clean water

CHECKPOINT 97

Improve welfare facilities and services together with workers.

WHY

Welfare facilities and services are an essential part of good working conditions. They include sanitary facilities, arrangements for drinking, eating and resting, and access to first aid, health care, transport and recreation.

Welfare facilities are more than just a legal requirement. They can greatly help to reduce fatigue, improve productivity and maintain workers' health. They must be improved and maintained by close cooperation of management and workers.

It is important that these essential welfare facilities and services serve their purpose. Planning and reviewing jointly with workers can ensure this.

HOW

1. Find out, through interviews, casual talk or simple questionnaires, what are the common needs of workers concerning welfare facilities and services.

2. Form a small joint team to study the existing situation and identify available options for improving welfare facilities and services. Involve workers' representatives, supervisors, and safety and health committee members or safety representatives.

3. Propose improvement plans starting with low-cost and easy-to-implement solutions. Present these plans to management and workers, and get their feedback. This feedback is important to identify practicable improvements.

4. Discuss jointly with workers or their representatives the steps to be taken for carrying out the proposed improvements. Evaluation of the improvements achieved, again involving workers or their representatives, should follow.

SOME MORE HINTS

- Welfare facilities are used every day in many ways. Using them can be difficult or easy, unpleasant or comfortable, a health risk or an aid to hygiene and better health. Extra joint efforts are certainly appreciated far beyond the time and money invested.

- There are a variety of low-cost ways to improve welfare facilities and services. Examples found in your own workplaces and in neighbouring enterprises will help. Try to learn from good examples which have been achieved locally.

- Repair and upkeep of welfare facilities is often neglected, but very important. Joint planning should also include repair and maintenance.

POINTS TO REMEMBER

Workers care about toilets, first-aid kits, lunchrooms and lockers. Make plans jointly with workers to improve essential welfare facilities now.

Figure 97a. An eating corner in a rest area

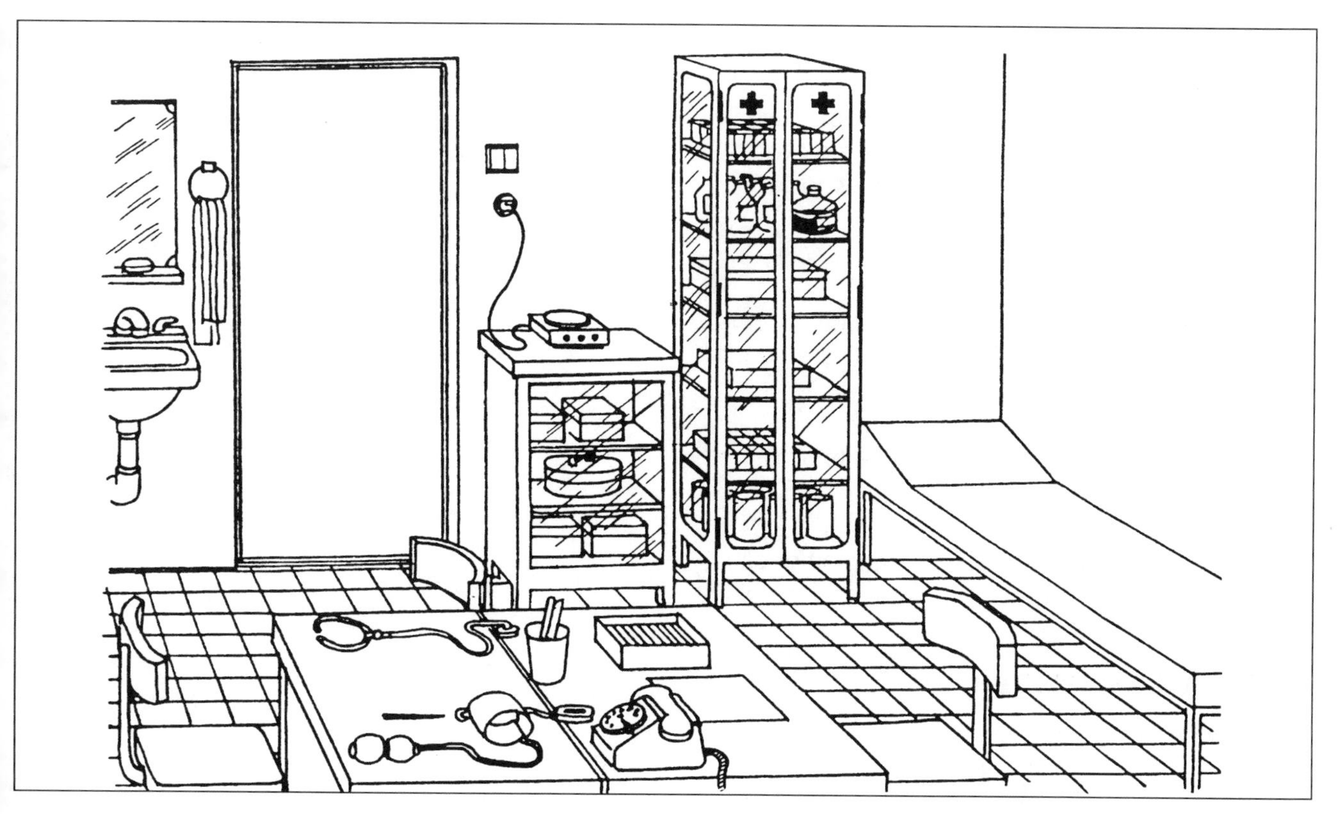

Figure 97b. A factory clinic served by a visiting doctor

CHECKPOINT 98

Provide a place for workers' meetings and training.

WHY

A good place for workers to meet and receive training will allow them to exchange ideas that are important to the enterprise's success and to the workers' safety and health.

Having an area set aside for meetings and training helps people to take their mind off work and allows them to focus on learning and problem solving.

HOW

1. Choose an area that helps to achieve the goals of the meeting or training. The area should be free from distractions and have a low noise level that allows easy listening.

2. Make sure that the place for meetings and training has enough furniture and adequate space so that people can be comfortable.

3. Provide good ventilation and lighting.

4. Control the use of this space so that meetings and training have priority. If it can be moved because of another event, your meeting or training will not seem so important.

SOME MORE HINTS

- Consider using a meeting-place where people want to be. A pleasant spot will motivate people to come and participate.

- Positive experiences in a good meeting-place make people eager to return.

- Enterprises which are close to each other can collaborate to arrange for a good meeting-place for their workers.

POINTS TO REMEMBER

If a meeting or training is important enough for workers to take time off from their work, then it should be held in a place where people feel that worthwhile things will happen.

Figure 98. Provide a place for meetings and training with enough space and furniture for workers to feel comfortable

CHECKPOINT 99

Clearly mark areas requiring the use of personal protective equipment.

WHY

Marking the areas requiring the use of protective equipment will help to create a habit of using the equipment. Marking the areas eradicates any doubts in the mind of workers as to whether personal protective equipment is needed or not.

Clearly marked areas where personal protective equipment should be used stress the need to do so. They make it easier for supervisors and workers to maintain the practice of using it without any ambiguity.

HOW

1. Identify work areas in which personal protective equipment is required to protect workers from specific hazards.

2. Obtain the type of personal protective equipment which is designed to protect against such workplace hazards and make it available to all workers who require it.

3. At each such work area, post signs with pictures which explain the type of personal protective equipment needed in that area.

4. Supervise and check the proper use of personal protective equipment in each of the designated areas. Organize regular inspection of these areas by a safety inspection team.

SOME MORE HINTS

- Manufacturers of personal protective equipment are the best source for obtaining signs.

- If a certain piece of equipment requires the use of personal protective equipment, try to install the sign on the equipment (for example, "EYE PROTECTION REQUIRED" on a grinding machine).

- Be sure that the personal protective equipment which is required in each work area is available in that area.

POINTS TO REMEMBER

Clearly marked areas requiring the use of personal protective equipment will help workers to remember to use their protective equipment all the time.

Figure 99. Clearly mark the areas where the use of particular personal protective equipment is obligatory

CHECKPOINT 100

Provide personal protective equipment that gives adequate protection.

WHY

Personal protective equipment which provides protection for a certain part of the body (e.g. hand protection or respiratory protection) comes in different types. Each type of personal protective equipment is designed to protect against certain hazards only.

It is imperative to match the personal protective equipment to each type of hazard which may be encountered in each work area.

Using the wrong kind of protective equipment gives the worker a false sense of security. This is very dangerous.

HOW

1. Identify the type of hazard in each work area.

2. Consult the manufacturers of personal protective equipment to ensure that you have the right kind of equipment for protection against the type of hazard in each work area. Designate one person or a team responsible for selection of personal protective equipment, and provide adequate training.

3. Provide a sufficient number of the right kind of personal protective equipment for each of the work areas requiring it.

4. Check regularly the proper use of the right kind of personal protective equipment.

SOME MORE HINTS

- When personal protective equipment is used for protection against chemical hazards, it is imperative to identify the chemicals.

- Although the general term "glove" is used to identify all types of hand protection equipment, that does not mean that any glove would provide protection against all chemicals. For example, a glove which is designed to protect hands against sodium hydroxide (caustic) may be inadequate for protection against solvents.

- There are no respirators that can protect workers from all chemicals. For example, an air purifying respirator which is designed to remove hydrogen sulphide from the air would be worthless against carbon monoxide. The worker using this type of respirator while working with carbon monoxide would have a false sense of security.

- In situations where oxygen may be limited (in a confined space), a filter-type respirator is dangerous. A unit which supplies breathing air to the worker becomes an essential part of the worker's protective equipment.

POINTS TO REMEMBER

Avoid using the wrong kind of personal protective equipment that would cause a false sense of security. Consult the manufacturer for any selection and use of personal protective equipment.

Figure 100a. Make sure that the chosen personal protective equipment gives adequate protection

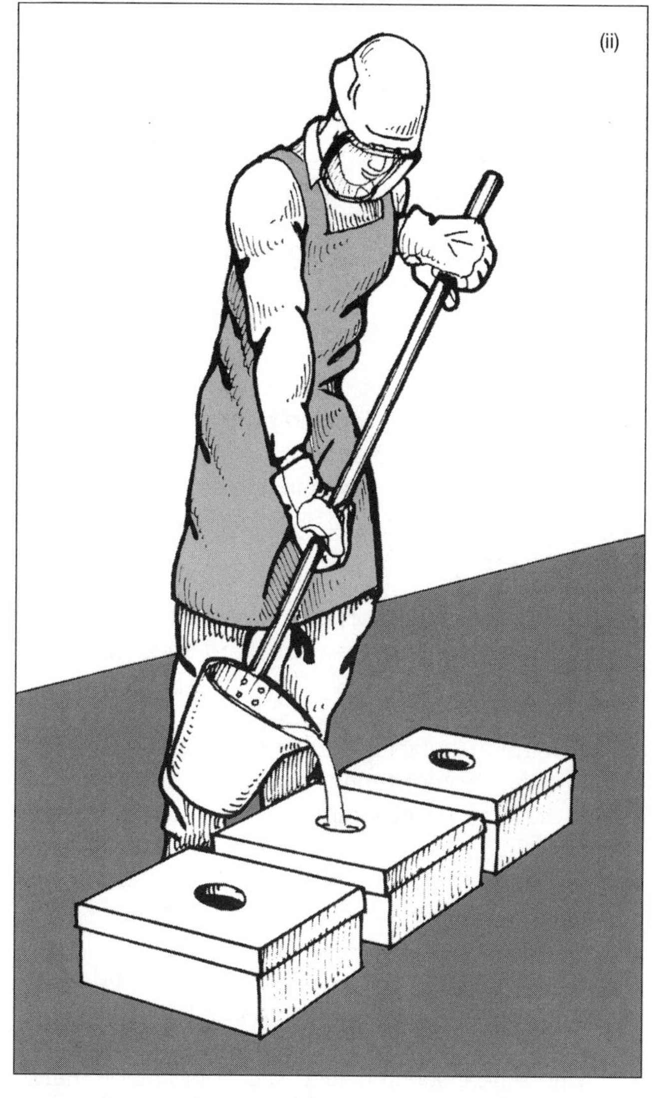

Figure 100b. (i) and (ii) It is extremely important to match the personal protective equipment to the particular type of hazard encountered in each work area. Check its proper use regularly

CHECKPOINT 101

Choose well-fitted and easy-to-maintain personal protective equipment when risks cannot be eliminated by other means.

WHY

Industrial work environments pose many hazards to workers' safety. Every attempt should be made to eliminate these hazards by changing work methods and modifying machines. For hazards which cannot be eliminated, proper personal protective equipment must be selected and used.

The selected protective equipment must be used properly. For this purpose, it is very important to select equipment that provides not only the best protection, but also good comfort and mobility for workers, as well as easy maintenance.

Ill-fitted protective equipment gives a false sense of safety. This is very dangerous and should be avoided.

HOW

1. The best way to select the proper personal protective equipment is to contact the manufacturer. Obtain sufficient information to select equipment which is acceptable to workers, easy to maintain and cost-effective.

2. It is extremely important to describe to the manufacturer the types of hazard encountered in your workplace.

3. Ask the manufacturer or its representative either to give a demonstration at your workplace or to provide you with samples prior to any purchase.

4. Compare the costs versus the degree of effectiveness.

5. Give high priority to workers' comfort and easy maintenance in the selection process. This is especially important in hot working conditions where heat accumulation and perspiration make the use of personal protective equipment difficult.

SOME MORE HINTS

- Bear in mind that workers show resistance towards using personal protective equipment which makes them uncomfortable. This problem exists in many workplaces.

- Often the discomfort of using personal protective equipment is combined with heat stress. Pay attention to preventing heat stress.

POINTS TO REMEMBER

In selecting personal protective equipment, workers' comfort and mobility, as well as easy maintenance, must be taken into account.

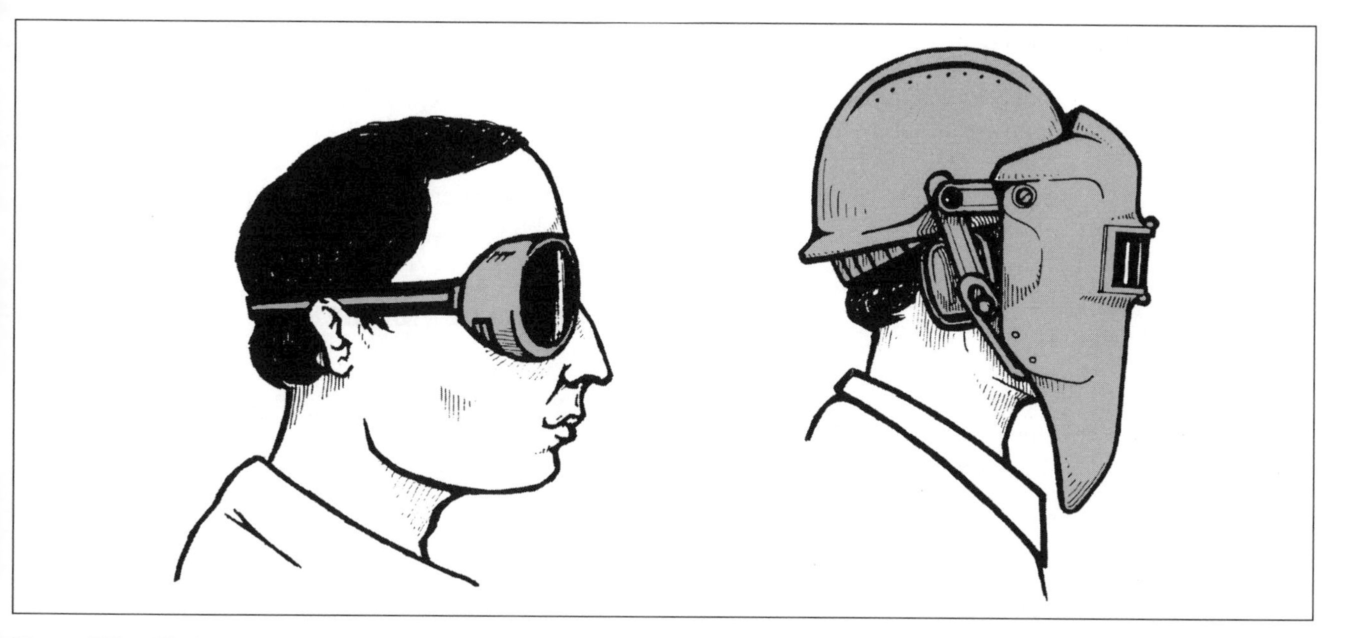

Figure 101a. Fitting the personal protective equipment to each individual worker is absolutely essential. Make sure that well-fitted equipment is provided and used

Figure 101b. Give high priority to workers' comfort and easy maintenance in selecting personal protective equipment from among available types

CHECKPOINT 102

Ensure regular use of personal protective equipment by proper instructions, adaptation trials and training.

WHY

When the use of protective equipment is essential, its regular use is vital as you can never foresee an accident happening.

Only regular use of personal protective equipment can effectively reduce exposure to hazardous conditions and protect workers in the long run.

HOW

1. Inform every worker requiring personal protective equipment, both by the spoken word and in writing, about:

   - why it is necessary to use the personal protective equipment;
   - when and where the personal protective equipment should be used;
   - how it should be used; and
   - how to care for the equipment.

2. Train workers sufficiently in the correct use and maintenance of their protective equipment.

3. Encourage workers to use their protective equipment for a trial adaptation period. Bear in mind that the user needs to have time to adapt to it during a supervised trial period of at least several weeks.

4. Supervise and check regularly the use and maintenance of protective equipment at work.

5. Provide spare parts and maintenance facilities at work for quick replacement of worn-out parts or equipment.

SOME MORE HINTS

- Regular use of personal protective equipment is ensured only when it is insisted on all the time and is checked regularly.
- Protective equipment makes an additional work demand. Encouragement and trials are always needed for its proper use.
- For respiratory protection, it is important to inform workers about what types of equipment and what types of filter should be used for their protection.

POINTS TO REMEMBER

Regular use of personal protective equipment at work saves money and reduces human suffering.

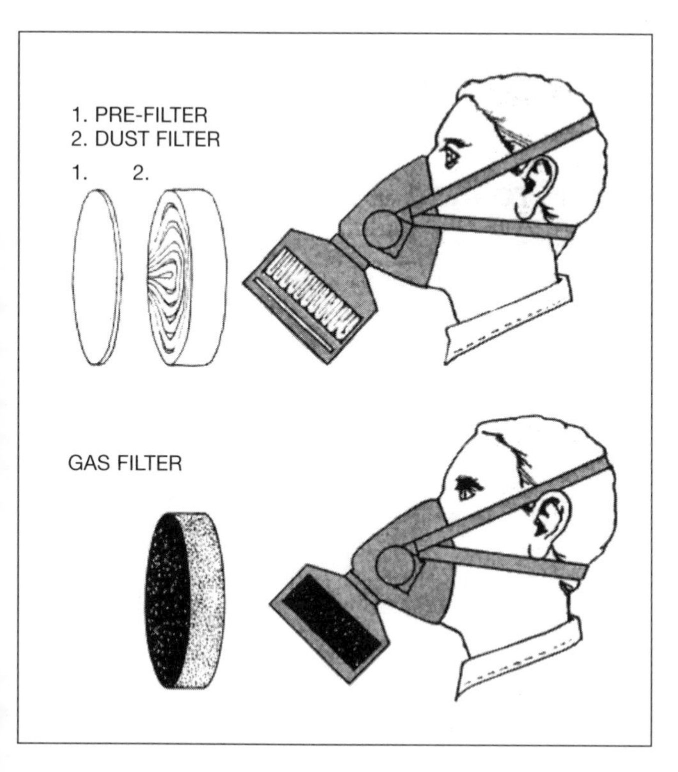

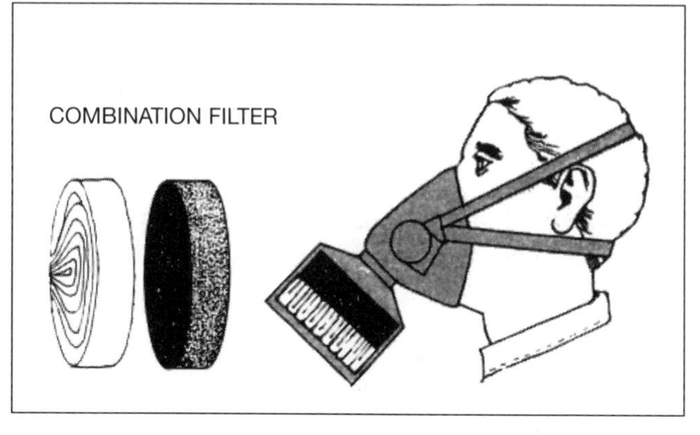

Figure 102a. Three types of half mask with filter. *Top left:* As protection against airborne particles, e.g. stone dust. *Bottom left:* As protection against gases and fumes, e.g. when using paints containing solvents. This filter contains activated carbon. *Above:* With a combination filter containing both a dust and a gas filter. These masks are examples of the simplest effective respiratory protection. Replace the filter when it gets harder to breathe or when it begins to smell. Replace the filter frequently

Figure 102b. All workers who may need to use respirators should be regularly trained in their use, care and maintenance

CHECKPOINT 103

Make sure that everyone uses personal protective equipment where it is needed.

WHY

Even the best personal protective equipment cannot protect workers against workplace hazards unless it is properly worn.

Workplace hazards do not cause death, injuries and illnesses every day. This gives workers the false assurance that personal protective equipment is not needed. A special training effort is essential.

HOW

1. Train workers on the hazards of their work environment.

2. Explain to the workers how the personal protective equipment can protect them against those hazards and how it fails to protect them when improperly used.

3. Remind the workers of the risks that they will be taking by not using their personal protective equipment.

4. Give encouragement for proper use of protective equipment and, where necessary, discipline workers who fail to use it.

5. Form a safety inspection team to walk regularly through different work areas and identify unsafe conditions, including situations where personal protective equipment is needed and not used.

SOME MORE HINTS

- Both managers and workers should identify workplace hazards and situations where personal protective equipment is required.

- To convince workers to use their protective equipment, it is essential to obtain the appropriate type of equipment.

- The safety inspection team must include both workers and management representatives.

- Be sure that the safety inspection team takes immediate corrective action and makes a written record of unsafe situations.

- Be aware that workers can develop a false sense of security. Make sure that they do not establish unsafe work habits because they rely on their protective equipment.

POINTS TO REMEMBER

Convince people to use their personal protective equipment properly at all times when it is needed. This needs consistent management.

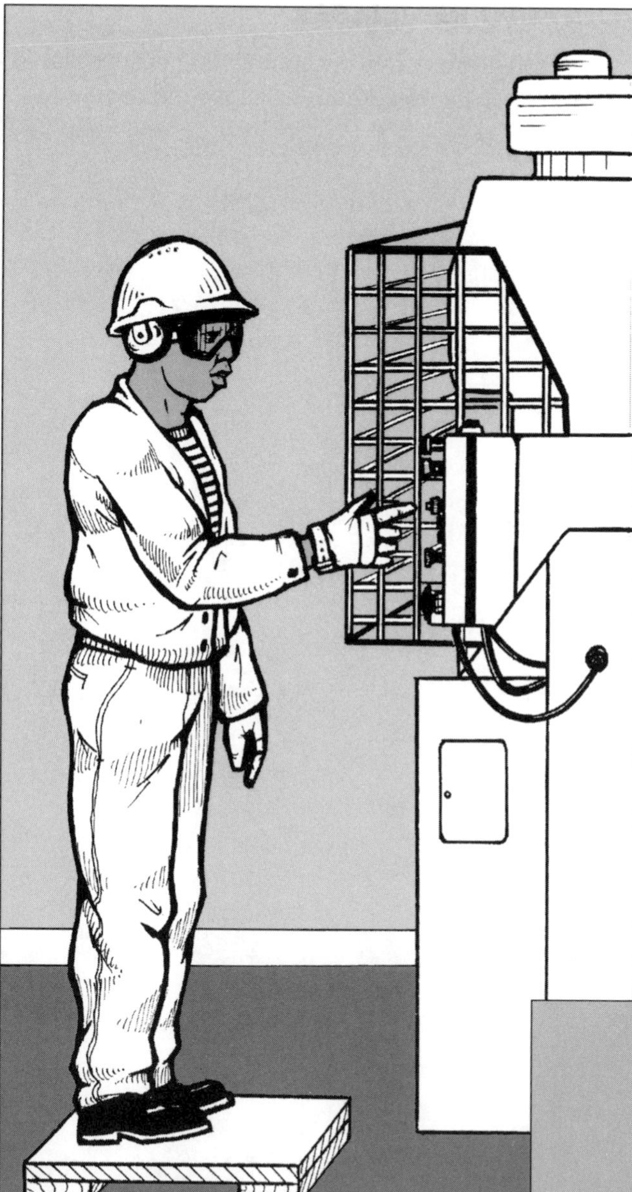

Figure 103a. Give encouragement for the proper use of personal protective equipment. This must be done in parallel with the enterprise effort to improve workplace conditions in general

Figure 103b. Both the manager or supervisor and the workers should identify workplace hazards and situations where personal protective equipment is required

CHECKPOINT 104

Make sure that personal protective equipment is acceptable to the workers.

WHY

If personal protective equipment is acceptable to the workers, its regular use is ensured, minimizing the risk of accidents and injuries.

Acceptable protective equipment reduces. tension and labour problems at the workplace.

There are requirements for acceptable personal protective equipment. Take sufficient care to ensure that the invested resources are properly used.

HOW

1. Provide not only the right kind of personal protective equipment, but also the type and size well fitted to each worker. Too tight or too loose equipment, for example, does not protect effectively, causes discomfort and discourages the user from regular use.

2. Provide the users of personal protective equipment with sufficient information about the risk factors at work and the equipment's potential for protection.

3. Make sure that everyone (supervisors, workers, worksite visitors, etc.) uses designated protective equipment where it is required.

4. Always select comfortable personal protective equipment, for example, lightweight equipment with proper ventilation and with maximum protection.

SOME MORE HINTS

- Adaptation trials before the regular use of personal protective equipment are helpful to convince workers that the equipment is necessary and acceptable.

- Consider workers' preferences regarding the colour, shape, material and design of the protective equipment.

POINTS TO REMEMBER

Personal protective equipment that is acceptable to the workers is used more willingly and regularly.

Figure 104a. Always select comfortable and well-fitted personal protective equipment

Figure 104b. Personal protective equipment is developing fast. Always select equipment which is effective and comfortable, for example, lightweight equipment with proper ventilation and with maximum protection

CHECKPOINT 105

Provide support for cleaning and maintaining personal protective equipment regularly.

WHY

The effectiveness of any personal protective equipment may be reduced with time and repeated use. Proper maintenance is essential for its regular use.

Clean and properly maintained personal protective equipment encourages the workers to use it regularly. Cleaning of the equipment should be incorporated within the maintenance programme.

Simply instructing workers to clean and maintain their protective equipment is not sufficient. Provide good and well-planned support so that they can easily cooperate in cleaning and maintenance of their own equipment.

HOW

1. Designate a group of persons responsible for the maintenance of personal protective equipment. Establish a good maintenance programme by consulting them and the workers concerned.

2. Identify how each kind of protective equipment should be stored, cleaned and maintained regularly. Make this known to all the workers using it.

3. Provide support for cleaning (e.g. providing good washing or cleaning facilities and, if necessary, assistance in the cleaning of clothing, etc.).

4. Provide support for maintenance and repair (e.g. by making it clear to whom the workers should address questions on the subject).

5. Make sure that all spare parts are available at all times.

SOME MORE HINTS

- When respirator filters must be changed regularly, assist workers in doing so by providing sufficient spare filters at clearly designated places. Regular checks should also be made by persons responsible for the maintenance programme of personal protective equipment.
- Each worker must use the protective equipment that is well fitted to his or her size. Make sure that this particular size is available when change or repair is needed.

POINTS TO REMEMBER

Cleaning and maintenance of personal protective equipment require a good maintenance programme. Designate a group of persons responsible for the programme.

Figure 105. The maintenance of personal protective equipment should be well planned, including storage, regular maintenance and training

CHECKPOINT 106

Provide proper storage for personal protective equipment.

WHY

Good management is the key to a sustainable programme for the use of any personal protective equipment. This should include a good storage policy.

Proper use and maintenance of personal protective equipment is facilitated by providing a "home" for each item.

Workers feel responsible for the appropriate use of protective equipment when storing it in a proper place after its use.

HOW

1. Check the number, size and quality of all the necessary personal protective equipment, and establish a policy on where and how to store each item.

2. In consultation with the users, designate an appropriate storage place for each item of protective equipment. Make sure that access to the equipment and its inventory are easy.

3. Make a concrete plan for regular checking of the use and maintenance of personal protective equipment. This will be facilitated by the fact that each item has its own place.

4. Maintain the storage place of personal protective equipment in good order.

5. Involve the users fully in all procedures from (1) to (4), above.

6. Good storage procedures should be an important part of the training programme concerning personal protective equipment.

SOME MORE HINTS

- Designation of storage places for personal protective equipment is best done by first establishing a good programme for the selection, use, maintenance, repair and review of such equipment.
- As storage places are fixed, it should be easy to provide adequate instructions about the use and maintenance of personal protective equipment and inform workers (e.g. by posting notices) of areas and processes where such equipment is required.

POINTS TO REMEMBER

Providing a "home" for each item is an essential part of the enterprise programme for the effective use of personal protective equipment. It represents the commitment of management and the workers concerned.

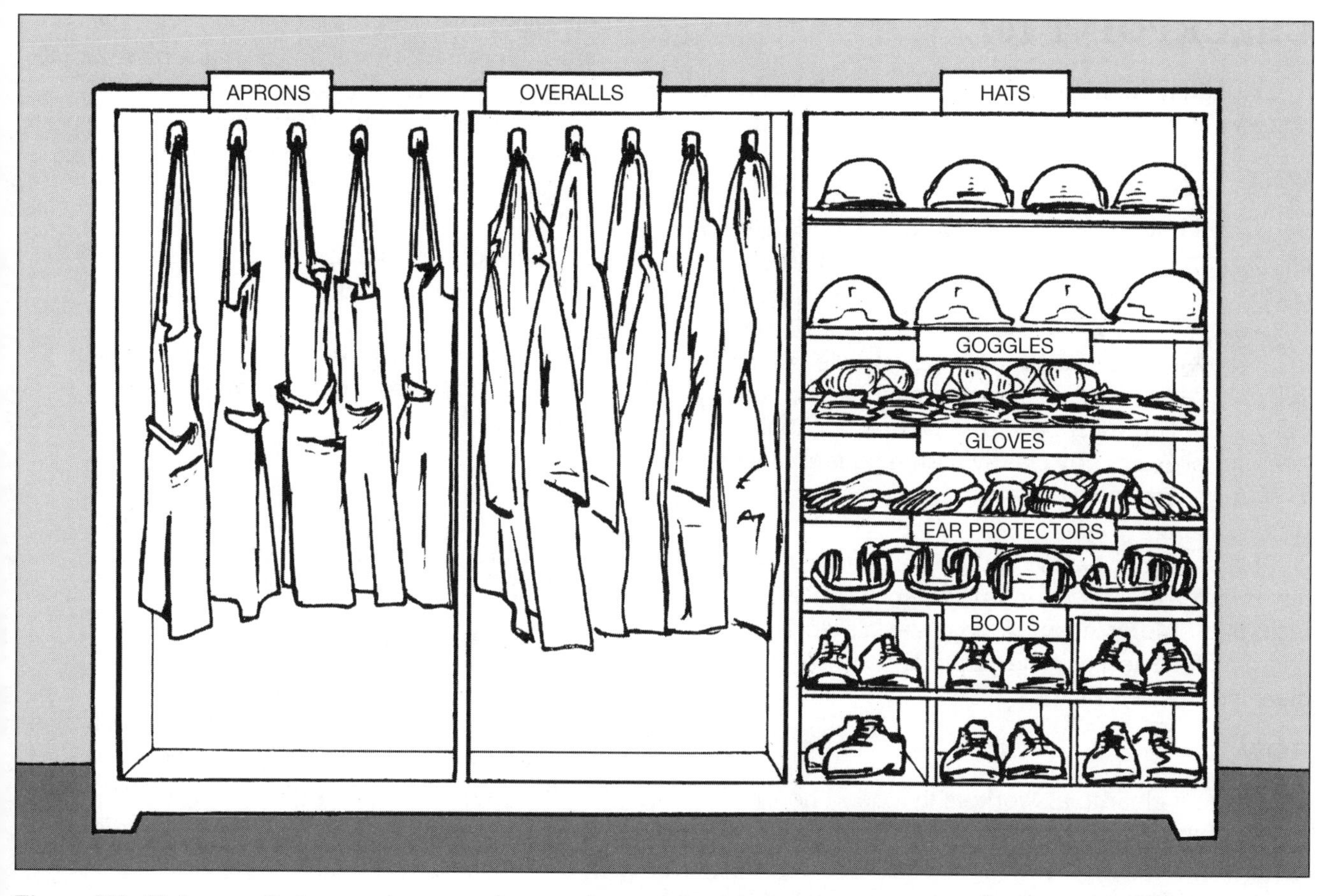

Figure 106. Make sure that access to personal protective equipment and its inventory are easy for workers and maintain the storage place in good order. Use cupboards with doors where possible as another way to keep personal protective equipment neat and clean

CHECKPOINT 107

Assign responsibility for day-to-day cleaning and housekeeping.

WHY

Good housekeeping cannot be left to spontaneous, unplanned activities. It needs planning and cooperation.

Good housekeeping for order and cleanliness will only begin when management takes responsibility for it. This must be shown by making housekeeping plans and by clearly assigning responsibility to supervisors and workers.

Day-to-day cleaning cannot be left until the last few minutes of the working day. Planning is essential, including orderly layout, good materials storage and handling facilities, a waste disposal policy and cleaning responsibilities.

HOW

1. Make it clear to all workers that good housekeeping is the enterprise's established policy and that therefore every effort is made to keep the premises orderly and clean, and to provide sufficient storage, transport and waste disposal facilities. (Checkpoints 1-21 in this manual related to materials storage and handling will help.)

2. Assign the responsibility to clean each work area to a particular group of workers with a leader or designated responsible person. This responsibility should include not only cleaning the area but also maintaining good housekeeping in the whole area.

3. Ask each such group to inspect the work area regularly and to evaluate the housekeeping performance.

4. Discuss with the group representatives what measures would be helpful to support their cleaning and housekeeping efforts.

SOME MORE HINTS

- Typical examples of measures needed to facilitate cleaning and housekeeping include marking of passageways and exits; setting aside special areas for storage; providing as many racks or stands as possible for materials and semi-products; using pushcarts and hand-trucks combined with the use of pallets; and providing waste receptacles.

- Use floor-covering materials suitable for the work and for cleaning.

- Provide designated places for storing cleaning equipment inside or near the work area to be cleaned.

- For dealing with hazardous chemicals during cleaning and housekeeping, special training about safety is always needed, taking into account the specific hazards involved.

POINTS TO REMEMBER

Good housekeeping needs good planning. The experience of planning and maintaining good housekeeping will be useful for organizing other workplace improvements. So start good housekeeping by clearly assigning responsibility for cleaning and housekeeping.

Figure 107. Assign responsibility for good housekeeping, cleaning and maintenance

Work organization

CHECKPOINT 108

Involve workers in planning their day-to-day work.

WHY

People enjoy their work more when they can control how they do it.

There are many ways that the work in your enterprise can be improved. The people who do the work know the most about these improvements. Involving people from the planning stage can uncover these useful innovations which might otherwise remain unknown to others.

Monotonous or repetitive work is made more difficult when people are not allowed to decide how the work is to be done. Planning together can improve this situation.

HOW

1. Examine how day-to-day work assignments are planned and to what extent workers themselves are involved in the planning. Then organize group discussions about how workers can be more actively involved in the planning process on a routine basis.

2. Where possible, allow workers to determine:

   - how fast the work is done (speed, cycle period);
   - in what order the work is done (timing, sequencing);
   - where the work is done;
   - who does the work.

3. Encourage people to present their ideas about ways of improving each work area. This can be done by having brief suggestion sessions or by organizing small-group discussions.

4. Set up autonomous groups in which workers can develop day-to-day work plans and the arrangements to do the work.

5. Keep a record of plans made and evaluate them regularly, also involving the workers.

SOME MORE HINTS

- Allowing people to control their work may not seem important to others, but may be very important to the person doing the work. This sense of self-control may also improve the work process.

- Controlling work features such as task assignments, pacing, priority and sequencing of individual work tasks may be especially important in the case of monotonous jobs.

- A group which is collectively responsible for quantity and quality of work is more productive and more disciplined than the same number of workers working completely separately.

POINTS TO REMEMBER

The effects of allowing workers to control how they do their work are clearly visible. This is a prerequisite of the successful enterprise of tomorrow.

Figure 108. Provide opportunities for workers to discuss their ideas about ways of improving each work area

CHECKPOINT 109

Consult workers on improving working-time arrangements.

WHY

Working-time arrangements can change even within the same length of working hours. There are a variety of arrangements that may differ: starting and finishing times; rest breaks; day-to-day differences in shift lengths; shift systems; flexible hours systems, etc. It is quite often necessary to seek better options.

When changing working-time arrangements, consulting the workers is the best way to develop better options.

New working-time arrangements affect all workers. Different workers may have different views. In order to overcome such differences, it is indispensable to involve everyone concerned from the planning stage.

HOW

1. Identify possible options for new working-time arrangements through group discussion that involves the workers concerned, or their representatives.

2. In so doing, consider that there are various ways of changing working-time arrangements. Common examples include:

   - change in starting/finishing times;
   - staggered hours;
   - inserting rest breaks;
   - averaging working hours over time;
   - allocating holidays;
   - flexitime;
   - shiftwork systems;
   - varying shift lengths;
   - part-time work;
   - job-sharing.

3. Compare possible options by knowing how both business requirements and workers' preferences can be accommodated. Then agree on concrete plans.

4. Get feedback from workers before having a test run or introducing new arrangements. Do not be in a hurry. Negotiation before implementation is always indispensable, and further adjustments are, as a rule, needed.

SOME MORE HINTS

- Both business requirements (operating time, staffing levels and production plans) and workers' preferences (changes in working hours, holidays, weekends, family responsibilities) must be duly taken into account. This needs careful planning through group study.

- It is often useful to establish a planning team that includes workers' representatives and supervisors. The team can try to identify practical options. The plans presented by the team can be used as a basis for further workplace consultations.

- Most working-time arrangements cover questions that need to be negotiated by collective bargaining. The possible options proposed by a planning team can certainly be used at this bargaining stage.

- Examples of working-time arrangements used in similar establishments can serve as workable models.

- It is usually preferable to introduce new working-time arrangements on a trial basis. Joint evaluation by management and the workers' representatives should follow.

POINTS TO REMEMBER

Working-time arrangements affect everyday life. Consulting the workers concerned gives better results and makes them more satisfied.

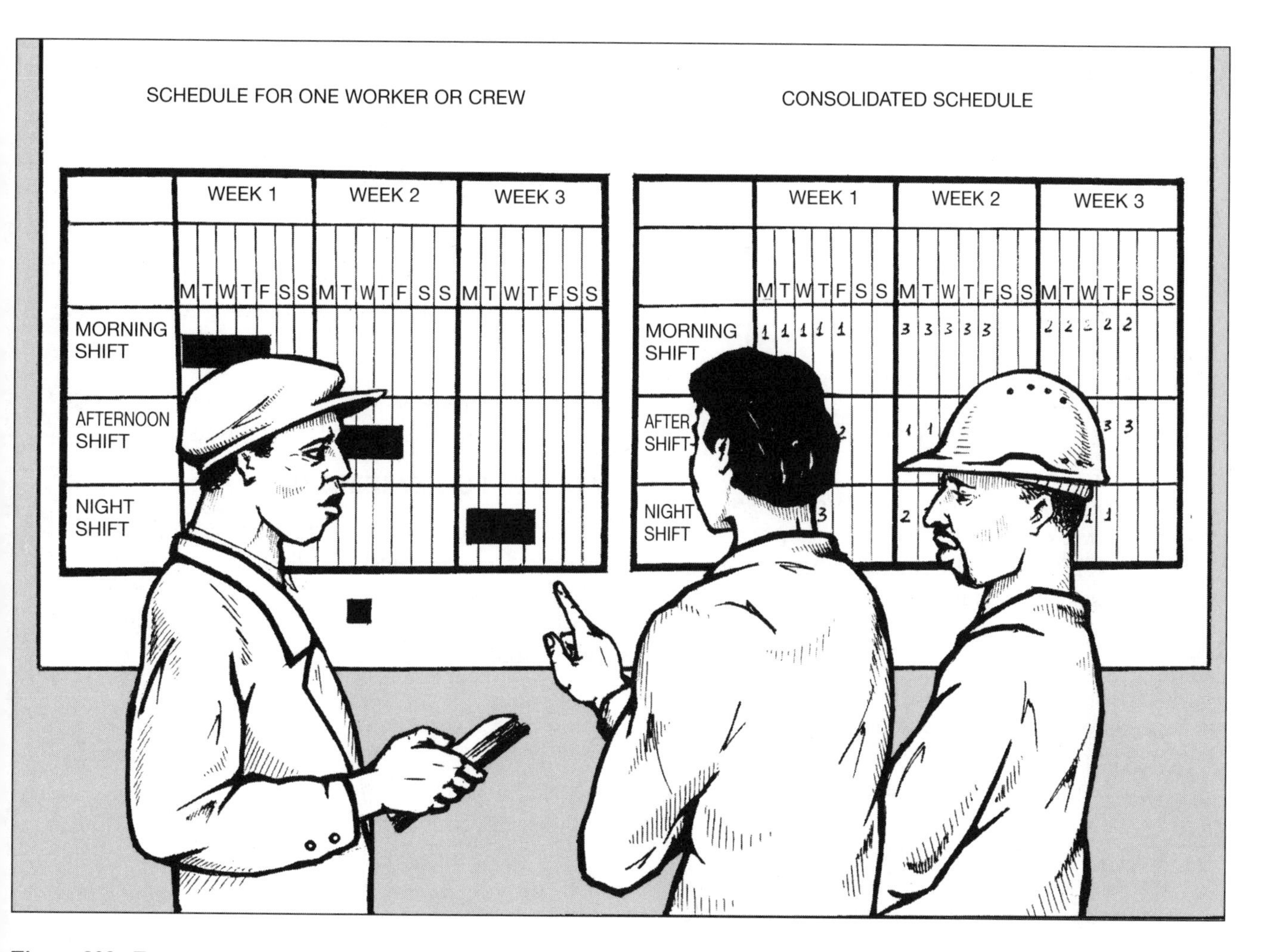

Figure 109. Encourage group discussion and get feedback from workers before introducing new working-time arrangements

CHECKPOINT 110

Solve work problems by involving workers in groups.

WHY

Workers, through their daily experience, know the origins of work problems and often useful hints about how to solve them, too.

Most work problems can be solved by relatively simple and inexpensive solutions. Group discussion is the best way to find such practical solutions.

Solving work problems often means new changes in working methods and work assignments. Workers involved in the planning of these changes will more easily accept them.

HOW

1. Consult workers about production bottlenecks and other work problems, and organize group discussion about why these problems have occurred and how they can be solved.

2. Form a small group (or several small groups depending on the size of the problem) and ask the group to come up with feasible options for solving the problem.

3. If technical advice is needed for the solution, provide adequate support for the group discussion in the form of information on practical improvements or experts' advice.

4. Present these options to all managers and workers concerned, and obtain their feedback. Select the most practical and effective solution based on the feedback.

5. Make known to all workers the proposals presented and the result of the implementation of improvements thus chosen. This encourages further promotion of participatory problem solving.

SOME MORE HINTS

- If there is a bottleneck operation or problem area in your workplace, it probably results from a combination of factors. Therefore it becomes necessary to take several actions at the same time. It is important to ask the worker groups to develop a practical set of solutions that can cover these several important aspects rather than one-sided solutions that leave out other, more important aspects.

- Make sure that workers know that they should report any problems and participate in solving them.

- Obtain the advice of someone who has experience in solving similar problems.

POINTS TO REMEMBER

Explain clearly the bottlenecks or other problems to workers and give them a chance to make suggestions. The best way to do so is to involve workers in group discussion about how to solve the problems.

Figure 110. Discuss in a small group (or several groups depending on the size of the problem in question) feasible options for solving the problem

CHECKPOINT 111

Consult workers when there are changes in production and when improvements are needed for safer, easier and more efficient work.

WHY

Workers will perform better in a new situation when they are involved in the process of change to that situation.

Workers' knowledge and experience help in solving production problems or in improving workplace conditions.

Many procedures, tasks and jobs are done in a certain way because they have always been done that way. There may be many better ways to achieve the enterprise's goals without much cost. These better ways can be more effectively found by involving workers who know the existing situation.

HOW

1. Ask workers which parts of the existing job are most difficult, dangerous and unpleasant, and how they think these problems can be solved.

2. Create an open environment where workers volunteer ideas for improving product design and work processes. This is usually done by having discussion sessions in small groups. It is necessary to show the workers that their ideas are responded to by quick action, or tell them immediately why it cannot be done.

3. When there are changes in product design or work processes, consult workers about such changes to find ways to make their work safer, easier and more efficient in the new situation.

4. Develop a procedure for receiving and acting on the workers' input (e.g. quality circles or planning committees).

5. Encourage and reward workers who present ideas for improvements.

SOME MORE HINTS

- Involve workers from the initial planning and design stages. This is much better than consulting them only after all essential plans have been made.

- Make plans for safe operations in emergency situations, too.

- Make the reward meaningful to the workers.

POINTS TO REMEMBER

Involving workers in making changes in product design or work processes makes them key to the success of their work.

Figure 111. Create an open environment in which workers feel free to examine risks and problems and exchange opinions about how to deal with them

CHECKPOINT 112

Reward workers for their help in improving productivity and the workplace.

WHY

Improving productivity and the workplace requires changes in the way work is carried out. This is effectively done by actively involving workers in planning and implementation of the improvement process.

It is important to show the commitment of the enterprise to constant improvement. Show this commitment by rewarding workers properly when they help in making improvements.

HOW

1. Make the improvement of productivity and the workplace as systematic a process as possible by encouraging suggestions from workers and by organizing group discussions involving workers (e.g. in the form of participatory group activities).

2. Establish a clear policy to reward workers who have helped in proposing or implementing practical solutions. Make known to all workers what kinds of reward they may expect to receive.

3. Reward these workers by appropriate means that suit the enterprise's overall policy. They can include announcing the best proposals or groups, giving awards, providing some forms of remuneration, inviting the best groups to special events or organizing ceremonial occasions.

SOME MORE HINTS

- Explain the reward system in company newsletters or in meetings during company time. Explain that both the company and workers can benefit from improvements achieved.

- Take action on workers' suggestions in a very obvious way, and be consistent in rewarding useful ideas and active participation.

POINTS TO REMEMBER

Make improvement of productivity and the workplace a systematic process by showing the enterprise's commitment and by properly rewarding workers who helped in the process.

Figure 112. Make it known that suggestions from workers are most welcome and organize group talks to discuss them. Reward workers who have helped in proposing or implementing practical solutions

CHECKPOINT 113

Inform workers frequently about the results of their work.

WHY

People learn and change by knowing exactly what other people feel and think about their work results.

Tell people if their work needs improvement so that they know what is expected of them. Also tell people when they are doing well. In that way, you can better communicate with each other and improve productivity.

Workers are often isolated from each other and do not have the opportunity to learn what happens after their part of the work has been done. Special care is needed to inform them about the results of their work.

HOW

1. Let people know that you appreciate their work when they do their jobs well. Be specific in telling them exactly what they did well.

2. When people are not doing their jobs well, tell them what they are doing wrong. Focus on what they are doing wrong and how to correct it, while also acknowledging their strengths.

3. Organize opportunities to show people how specific jobs should be done through examples and demonstrations by other experienced workers.

4. Check whether people are told regularly about the results of their work. Keep in mind that this should be done in such a way as to avoid giving the impression that the work is being supervised for strict disciplinary purposes. Tell workers about their work results to let them know how important this is to the workers, other people and the enterprise as a whole.

SOME MORE HINTS

- When people know that they are doing their job well, they develop a sense of pride and self-worth. This allows them to be even better workers in the future.
- It is true that people are afraid of criticism. But we can tell people that they are doing their jobs incorrectly, not for the sake of criticism but in order to work together better. This sense of working together should be conveyed by making it a rule within the enterprise to inform people regularly about their work results in a friendly manner.

POINTS TO REMEMBER

People want to do good work. By telling them about how they are doing, you can help them achieve this goal.

Figure 113. Organize opportunities to let workers know how specific jobs should be done through examples and demonstration of good or bad work results

CHECKPOINT 114

Train workers to take responsibility and give them the means for making improvements in their jobs.

WHY

Interesting and productive jobs are those in which workers take responsibility for planning and output. Responsible jobs can lead to increased job satisfaction.

Jobs which have no real responsibilities are not only boring but require continuous supervision, and thus become burdensome both for the enterprise and the workers.

We all need to feel that our work has some value and that we can develop our abilities and skills. Towards this end, it is important to train workers to take on responsible jobs.

HOW

1. Organize group discussions on how to improve jobs. Include in the discussions ways in which jobs with more responsibilities can benefit both the enterprise and the workers.

2. Incorporate discussions on work organization and job content within training sessions on job improvement and career development.

3. Use examples of well-organized jobs that can increase job satisfaction in such training.

4. Promote arrangements for group work as this can increase the awareness that more responsible jobs for the group are more interesting and good for skills development.

5. Provide good training opportunities, either on the job or through special training sessions or courses, for taking on more responsible jobs with multiple skills.

SOME MORE HINTS

- Increase mobility within the enterprise so that the same worker can be assigned to different kinds of job and can thus learn to take responsibility in different situations.

- Make sure that taking more responsibility at work can lead to better work results and can be rewarded better in the long run.

- Discuss with workers the jobs in your enterprise which combine appropriate responsibilities and are productive.

POINTS TO REMEMBER

By taking on more responsibility, the worker can see the connection between his or her own work role and the overall activity of the enterprise. This makes the work more productive and more satisfactory in the long run.

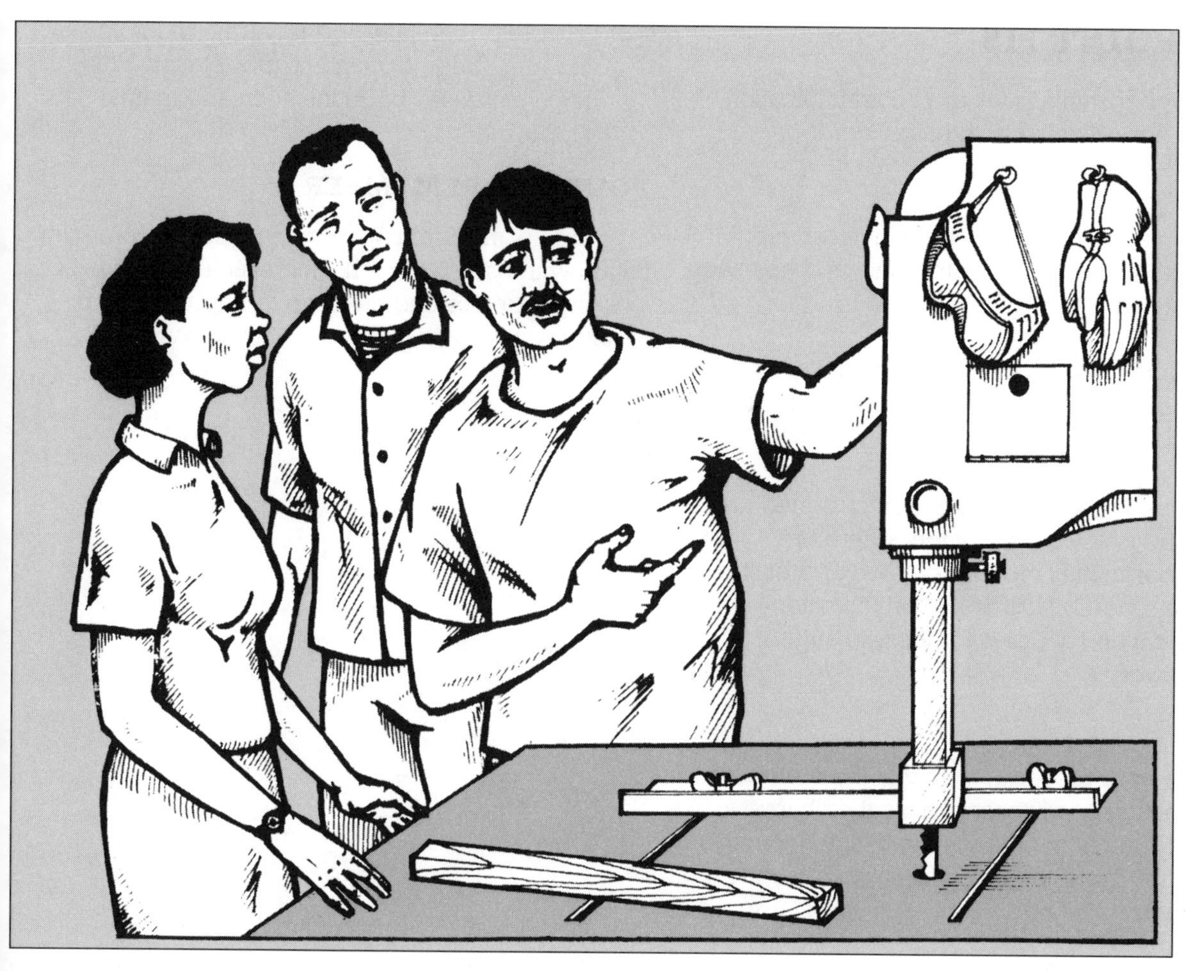

Figure 114a. Train workers to take on more responsible and safer jobs

Figure 114b. Make the company plan for better workplaces known to each individual worker and encourage mutual communication

CHECKPOINT 115

Provide opportunities for easy communication and mutual support at the workplace.

WHY

Jobs are carried out far better when people know what others are doing or thinking, and how they can cooperate with each other.

Poor communication often leads to delays in work or low quality of products, and even to mistakes and accidents.

People are busy completing their assigned tasks and tend to be isolated from others. Therefore concrete opportunities must be created and built into daily work in order to facilitate communication among workers and in order for them to support each other's work.

HOW

1. Arrange work procedures so that the members of the section or work team have the chance to communicate from time to time. Also encourage casual talk. Avoid totally isolated work as much as possible.

2. Organize brief meetings, if appropriate before every shift, to deliver instructions, exchange the day's work plans and have a question-and-answer session.

3. Encourage group planning and group implementation of tasks, especially by assigning work to the group instead of to individuals. This facilitates communication and constant cooperation.

4. Provide adequate opportunities for training and retraining workers within daily work. This helps to improve communication and mutual support.

SOME MORE HINTS

- Use newsletters, leaflets, updated instructions, posters and occasional verbal presentations to increase communication.

- Provide changing-rooms, rest areas, drinking facilities and eating areas for joint use so as to give workers more chance to talk to each other.

- Provide possibilities for acquiring multiple skills and encourage occasional job rotation. This helps increase communication and mutual support.

POINTS TO REMEMBER

Create, intentionally, more chances of communicating with each other. This increases the sense of working together and can lead to improved work results.

Figure 115a. Encourage communication at work and group implementation of tasks

Figure 115b. Assign work responsibilities to a group of workers instead of individuals. This helps to increase mutual communication and thus facilitates better work flows and improved work results

Figure 115c. Organize training sessions within daily work

CHECKPOINT 116

Provide opportunities for workers to learn new skills.

WHY

Work methods are rapidly changing with the introduction of new technologies. By training workers in new skills, it is easier to organize new work systems which are more productive and safer.

By acquiring new skills, workers can do multiple jobs. This greatly helps the organization of job rotation and the replacement of absent workers without looking for additional workers.

Multiple-skilled workers can more easily set up group work to improve efficiency and to cut supervisory costs.

HOW

1. List new skills which workers need and wish to learn. Check how the opportunities for learning these new skills can be provided; on the job, through special training sessions or by sending selected workers to training courses held outside.

2. Encourage all workers to learn new skills by making it known to them in writing what opportunities exist and how they can apply for them.

3. Make plans for learning new skills by asking workers to propose practicable plans and discussing these proposals with them.

4. Try to organize the required training within working hours.

5. Try to introduce group work arrangements so that workers have real opportunities to use the new skills that they have learned, for example, by exchanging tasks or sharing work.

SOME MORE HINTS

- If necessary, organize special short training sessions on new skills to confirm the training needs and to encourage people to participate in further training.

- Evaluate progress in learning new skills regularly (e.g. once a year) and improve the arrangements further.

- Make full use of training courses offered by training and other institutions.

POINTS TO REMEMBER

Workers with multiple skills can flexibly help to overcome bottlenecks by exchanging tasks, sharing work and setting up productive group work arrangements.

Figure 116. Have plans for learning new skills based on workers' own suggestions

CHECKPOINT 117

Set up work groups, each of which collectively carries out work and is responsible for its results.

WHY

Today, many enterprises find it beneficial to assign work to groups instead of to individuals. This is because work groups are more productive, with much less unnecessary work and fewer mistakes.

Using group work arrangements, it is easier and less time-consuming to set tasks for a group than for individuals. As a result, less supervision is needed and the work flows more smoothly.

In group work, workers have better opportunities for communication and acquiring multiple skills.

In group work, workers can flexibly help each other to overcome bottlenecks and become collectively responsible for productivity, quality and discipline. This helps to create a good work climate.

HOW

1. Assign the responsibility to plan and implement a sequence of tasks to a group.

2. Consider grouping assembly or similar workers around one table and increase arrangements by which the workers help each other and share the tasks.

3. Replace a rigid conveyer line through "group workstations" with buffer stocks between them.

4. Introduce mechanized or automated processes in such a way that a group of workers using the processes work together in planning and day-to-day operation.

5. Train workers to acquire multiple skills so that they can exchange tasks and share work within the work groups.

6. Make sure that rewards depend on the performance of the group as a whole and not on the performance of individual group members.

SOME MORE HINTS

- Make sure that each work group can obtain the information and expertise it needs, for example, concerning supplies and maintenance.

- The group should have control over the methods used to do the work and over the way the work is shared among group members.

- The group should be given regular information about its performance that can be shared by all the group members.

- Make sure that there is no "outsider" in any of the groups. Disruption by outsiders who feel no responsibility for the group's work can cause a lot of problems.

POINTS TO REMEMBER

Autonomous work groups that are collectively responsible for work planning, the way work is shared and product quality are very productive since groups can work faster and better than the same number of separate individuals.

Figure 117. Assign to a group the responsibility to plan and implement a sequence of tasks

CHECKPOINT 118

Improve jobs that are difficult and disliked in order to increase productivity in the long run.

WHY

In every enterprise, there are bottleneck operations which are particularly difficult and therefore disliked by workers. Special effort is needed to improve these bottlenecks.

Until recently, it was assumed that job characteristics were predetermined by technical and economic requirements. However, it is today possible to design better jobs by using more up-to-date technologies and improved work organization. There is good scope for defeating difficult and monotonous jobs.

By improving difficult jobs, it becomes easier to assign jobs, rotate workers and make effective production plans.

HOW

1. Examine the jobs within the enterprise that are considered to be difficult. Typical examples of such jobs are:

   - physically demanding tasks, such as manual handling of heavy materials;
   - work exposed to excessive heat, cold, dust, noise or other hazardous agents;
   - jobs that are often performed in irregular working hours such as frequent night shifts;
   - repetitive work that is fragmented, boring and isolated;
   - jobs requiring little skill with only limited career possibilities;
   - skilled but arduous jobs that are stressful and fatigue workers.

2. Mechanize difficult tasks but avoid jobs resulting in machine-paced or monotonous tasks.

3. Improve equipment and sequence of work to make the job easier and more responsible.

4. Combine tasks (e.g. to constitute a job performing a sequence of assembly tasks) so as to have a longer cycle time.

5. Make the job less machine paced or conveyor paced (e.g. by having both upstream and downstream buffer stocks of unfinished products which allow the worker to take a pause or change the pace of work).

6. Add more responsible tasks such as inspection, maintenance and repair.

7. Promote multi-skilled jobs and sharing of jobs to avoid concentrating difficult jobs on a small number of workers.

SOME MORE HINTS

- A very flexible way of improving job content is group work. This makes possible the overlapping of skills and the sharing of difficult work.
- Elimination of difficult jobs must also be planned by involving workers. Group discussion of the change process is indispensable.
- Emphasize that the reduction of difficult jobs not only reduces occupational stress and ill-effects, but also facilitates better use of skills and improves career prospects. The benefits will include a more productive enterprise.

POINTS TO REMEMBER

As there are no simple solutions to the problems of difficult jobs, take advantage of suggestions from both managers and workers. Improvement is usually needed in equipment and work methods, as well as in work organization.

BEFORE

AFTER

Figure 118a. (i) and (ii) Improvement is needed not only in work equipment but also in the way work is organized. Group work to do a sequence of jobs can be a good starting-point for improving work organization

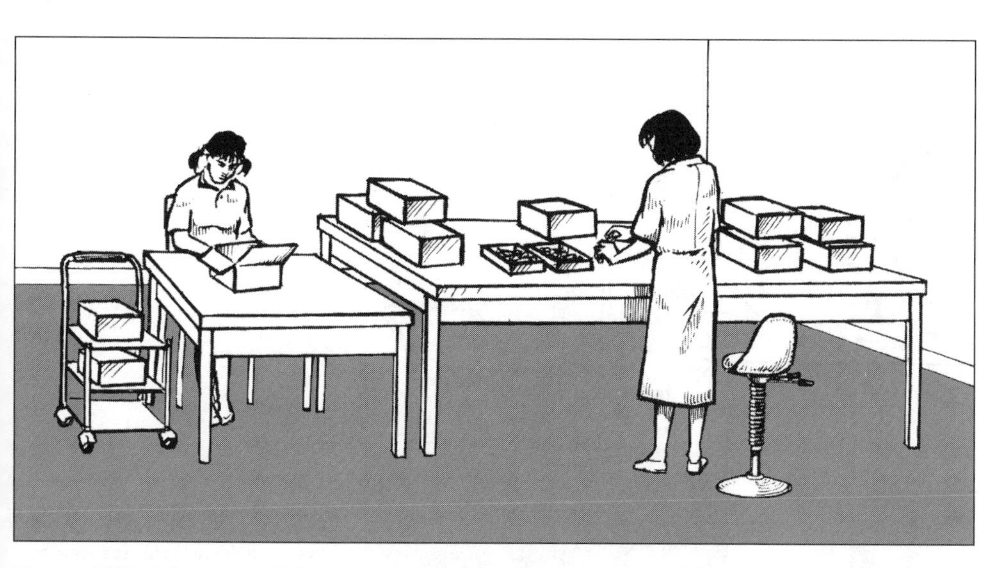

Figure 118b. There are different ways of doing the same job. By improving difficult or uninteresting jobs, it becomes easier to assign jobs, rotate workers and make more effective work plans

CHECKPOINT 119

Combine tasks to make the work more interesting and varied.

WHY

Repetition of the same monotonous tasks and lack of variety cause boredom and fatigue. The result is low efficiency and a negative work attitude. Frequent changes in tasks are needed.

Monotony can cause attention to wander. This can easily lead to low-quality work and even to accidents. Monotony must be defeated to keep workers alert and productive.

Performing a number of tasks prepares workers for multiple skills. Multi-skilled workers are more productive and help the enterprise to organize a better flow of work.

HOW

1. Combine two or more tasks to be done by one worker. Provide the necessary changes in workstation and tools.

2. Combine a series of tasks so that the cycle time per worker becomes longer.

3. Allow rotation of jobs within a certain number of workers so that each worker can have frequent changes of task.

4. Arrange for autonomous work groups in each of which several workers have joint responsibility for performing the combined tasks and share the work.

5. Train workers for adequately performing new, combined tasks.

SOME MORE HINTS

- Provide workstations which the same worker can use for performing multiple tasks and which can thus be used by different workers.

- While combining tasks, provide opportunities to walk around or change from sitting to standing or from standing to sitting.

POINTS TO REMEMBER

Combine tasks to defeat monotony and make the work more productive.

Figure 119a. Provide multi-task workstations for use by different workers. This helps the enterprise organize a better flow of work

Figure 119b. Combine two or more tasks to be done by one worker so that the cycle time is longer and the work more interesting

CHECKPOINT 120

Set up a small stock of unfinished products (buffer stock) between different workstations.

WHY

Small supplies of unfinished workpieces in front of and behind each workstation (so-called "buffer stock") eliminate time spent waiting for the next workpiece. They also help to eliminate time pressure as the next worker or machine will not have to wait either.

Working at the worker's own pace without time pressure gives much flexibility at work. It also gives the sense of being independent, treated fairly and better organized. This can lead to improved productivity in the long run.

Such buffer stock is part of a modern concept that machine-paced tasks, such as conveyer belt work, should be replaced by more flexible work organization.

HOW

1. Rearrange the work flow so that there can be a small stock of unfinished products (buffer stock) between subsequent workstations (e.g. between workstations A and B, between workstations B and C, between workstations C and D, etc., when the work flow is from A to B, B to C, C to D, etc.).

2. Set up places for this small stock of unfinished workpieces, considering the size, type and possible number of workpieces that may be placed there.

3. In the case of small workpieces, simple bins or small pallets with dividers are usually sufficient.

4. For bigger and heavier workpieces, like assembled metal products or large wooden items, special racks or pallets, or mobile storage shelves should be used.

5. Minimize the floor space taken up by the buffer stock, and ensure easy access by the next worker.

SOME MORE HINTS

- Choose the appropriate height for the buffer and design it so as to minimize the effort needed to put stock in or take it out.

- Stock workpieces in a systematic manner so that they can be seen at a glance and so that their handling is easy.

- When buffer stocks are present, workers can build up a small advance which they can use to take a few seconds' rest, correct machine settings or fetch some spare parts without slowing down the operation as a whole. This assures continuity and flexibility.

- If transport of the buffer stock from one workstation to the next process is necessary, it is useful to provide a mobile rack to store the buffer after finishing the work at this workstation.

POINTS TO REMEMBER

Buffer stock (small supplies of workpieces between workstations) is used in many modern production systems. This is a symbol of good work organization.

Figure 120a. Assembly line with intermediate buffer stocks

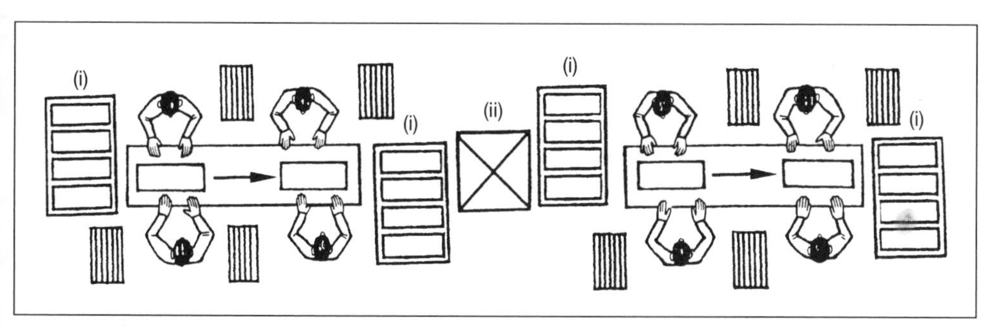

Figure 120b. New arrangements based on group workstations and buffer stocks. (i) Buffer stock. (ii) Automated assembly unit. Note that the buffer stocks and group workstations allow for partial automation without disruption of the production process

Figure 120c. Table-top rotatable buffer line

CHECKPOINT 121

Combine visual display work with other tasks to increase productivity and reduce fatigue.

WHY

Work at a visual display unit (VDU) requires a fixed posture and strains the eyes. The resulting pain and discomfort can be prevented by combining VDU work with other tasks, in addition to providing an adjustable table and chair.

Variety in work tasks can increase job satisfaction, leading to improved well-being and higher productivity.

Prolonged VDU work is usually connected with mere repetition of simple tasks (e.g. data entry work). Mixing these simple tasks with more skilled work and non-VDU work can achieve variations in physical, visual and mental demands, and thus reduce common problems reported from VDU work.

HOW

1. Combine repetitive elements of work with non-repetitive, dialogue-style tasks. For example, combine data entry tasks with data dialogue and data inquiry tasks.

2. Organize work for a group of operators together so that each operator can perform multiple tasks.

3. Rotate jobs so that each worker can perform both VDU tasks and non-VDU tasks (such as conventional office work or other activities).

4. Retrain your workers in both VDU and non-VDU tasks. This will improve flexibility in work organization and result in better utilization of equipment and human resources, as well as improving workers' morale.

SOME MORE HINTS

- Encourage workers' participation in finding non-VDU tasks that they would like to combine with their regular VDU work.

- Promote training of VDU workers in more skilled VDU work (which normally involves varied tasks and dialogue work) and in non-VDU work available at the enterprise.

POINTS TO REMEMBER

Workers who combine VDU and non-VDU tasks during the working day are generally more satisfied and report fewer complaints.

Figure 121. (i) and (ii) Combine work at a visual display unit with other tasks so that continuous visual display work is avoided

CHECKPOINT 122

Provide short and frequent pauses during continuous visual display work.

WHY

Visual display work tends to keep the worker in a fixed posture and strains the eyes. Rest pauses help to maintain performance by preventing the onset of fatigue.

Prolonged VDU work increases mistakes. Short rest pauses can recover attention and concentration, resulting in improved work quality.

Taking short breaks at relatively short intervals (say, every hour) is better than taking a long break after the worker reaches a stage of excessive fatigue.

HOW

1. Allow for short breaks after, say, every hour of work. Work without a pause (e.g. two to four hours without a break) is not advisable. Changing body positions and focusing the eyes on something other than the screen will reduce fatigue.

2. Allow for the insertion of short tasks which are different from VDU work. For example, changing seated positions, switching to standing work, or taking a short walk to fetch something or for communication, greatly help to reduce muscle and eye fatigue.

3. Spend the break period away from the VDU workstation.

SOME MORE HINTS

- Resting your eyes occasionally away from the screen is necessary to prevent eye fatigue. This is difficult to do unless you leave the VDU workstation. Therefore pauses help.

- Combine your pauses with relaxing exercises, such as walking, stretching or simple gymnastics.

- Taking pauses because you are fatigued is less effective than taking pauses prior to the onset of fatigue. Therefore make it a rule to have a pause at regular intervals, say every hour.

POINTS TO REMEMBER

Take frequent short pauses so that your body and mind can gain renewed power.

Figure 122a. Provide short and frequent pauses during continuous visual display work

Figure 122b. Combine your pauses with relaxing movements

CHECKPOINT 123

Consider workers' skills and preferences in assigning people to jobs.

WHY

Workers are different from each other. Their skills and strengths differ and their preferences also differ. Some workers may be overstrained, while others are underutilized. Finding the proper jobs for these different workers needs constant planning and review.

Poorly assigned jobs can mean many lost opportunities and extra costs. Careful assignments offer many benefits.

Workers' preferences are as important as their capacities and skills. Take them into account to motivate workers and to help them feel individually responsible for their work.

HOW

1. Know each worker's skills and preferences, and consult the worker and those experienced in job design about the job to be assigned.

2. In assigning jobs, first consider if the jobs are well designed according to the following principles:

   - jobs should make clear who is responsible for output and quality;
   - jobs should help workers to develop skills and become interchangeable;
   - jobs should occupy each worker fully but should remain within each worker's capacity.

3. Combine tasks so that each worker is responsible for a good set of tasks that becomes interesting and requires trained skills.

4. Assign each worker to a job that is best suited to the worker's skills and preferences.

5. Provide training and retraining as required to improve job assignments.

SOME MORE HINTS

- Consider that if you do not combine tasks, it is very difficult to keep workers fully occupied. Fragmented tasks do not attract workers and therefore make it difficult to meet their preferences.

- Proper job assignments cannot be achieved by simply selecting workers for each existing job. Efforts are always necessary to improve the way existing jobs are done.

- Good job assignments can reduce the cost of supervision and make workers responsible for the output and quality of the work performed in the job.

POINTS TO REMEMBER

Assign to each worker a responsible job that best suits his or her skills and preferences. This is achieved by combining knowledge of the worker and improvement of the job design.

Figure 123. In assigning people to jobs, take into account not only workers' skills but also their preferences

CHECKPOINT 124

Adapt facilities and equipment to disabled workers so that they can do their jobs safely and efficiently.

WHY

Disabled workers can work safely and efficiently if adequate support is provided to meet their needs.

The needs of disabled workers are individually different. Some needs can be met by making the equipment and tasks more user-friendly, but there are other individual needs that can be addressed by paying close attention to these needs.

The best way to meet these needs is to organize group discussion about how the workplace can be improved and what are the priorities.

HOW

1. Organize group discussion about how to meet the special needs of disabled workers. Keep in mind that user-friendly measures can generally help, but that there are also individual needs to be considered.

2. Check various aspects of work, using this manual, to see what improvements at the workplace can solve the problems of disabled workers.

3. Consider not only easy access and use of work equipment but also easy access and use of equipment and facilities for workers' general and daily needs, such as movement of people, general instructions and welfare facilities.

4. Organize adequate training about meeting the needs of disabled workers not only for disabled workers themselves but also for all managers and workers.

SOME MORE HINTS

- Interview both disabled workers and other workers about how to meet the needs of disabled workers. The results can be used for group discussions.

- Flexible work organization is needed particularly for disabled workers. Discuss possible options in group meetings to find a practical solution.

- Learn from good examples in your own and other workplaces. Discuss these examples to find out if similar arrangements can be applied.

POINTS TO REMEMBER

By providing adequate support, disabled workers can work safely and efficiently. Organize group discussion involving them and other workers.

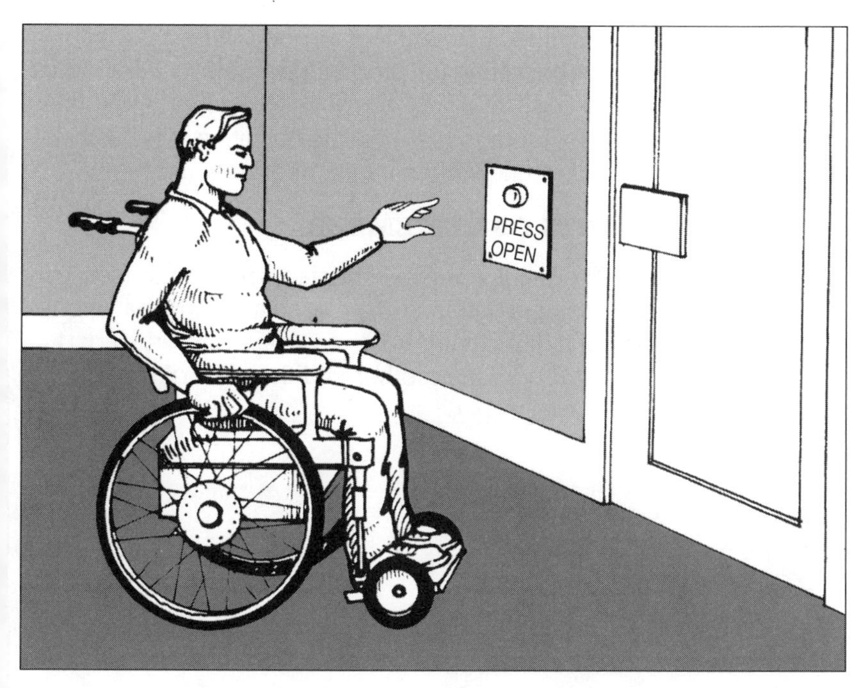

Figure 124a. Consider the easy access and use of equipment and facilities for disabled workers, taking account of their particular needs

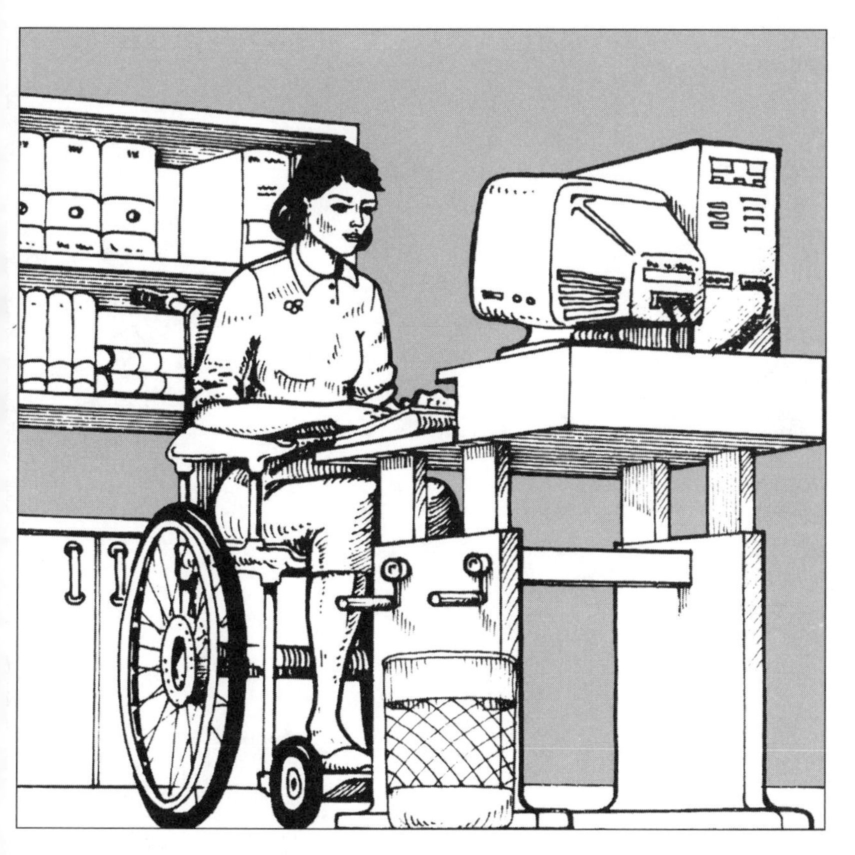

Figure 124b. By providing adequate support, disabled workers can work safely and efficiently

CHECKPOINT 125

Give due attention to the safety and health of pregnant women.

WHY

The conditions at work should not put the pregnant woman and her unborn child at risk. As women actively participate in all occupations, it is important to make sure that the safety and health of pregnant women are given due attention.

The physical condition of a pregnant woman at the later stages of pregnancy requires special attention. In particular, physically demanding tasks and arduous work, such as night work, must be avoided.

HOW

1. Do not assign pregnant women, especially during the last months of pregnancy, to lifting tasks, transport of loads and other heavy manual tasks.

2. Do not assign pregnant women to night work or other arduous tasks during the last months of pregnancy.

3. Make sure that access and space for movement around machines and equipment, and between workstations, are sufficient to allow easy and comfortable movement of pregnant women.

4. Provide sitting facilities for pregnant women. Do not assign pregnant women to tasks requiring prolonged standing or sitting.

5. Make the tasks assigned to pregnant women, especially in the last months of pregnancy, flexible enough so that they can take adequate rest pauses at work. If necessary, arrange for job rotation so that pregnant women can have a self-regulated pace of work.

SOME MORE HINTS

- Where possible, assign pregnant workers to sedentary tasks that are not physically demanding. It is important that pregnant women are not obliged to keep the same work posture all the time.

- Provide sufficient rest periods during the working day for pregnant women.

- Provide adequate welfare facilities at work that pregnant women can comfortably use.

POINTS TO REMEMBER

Pregnant women, especially during the last months of pregnancy, have special needs that should be considered in order to ensure the safety and health of both the mother and the unborn child.

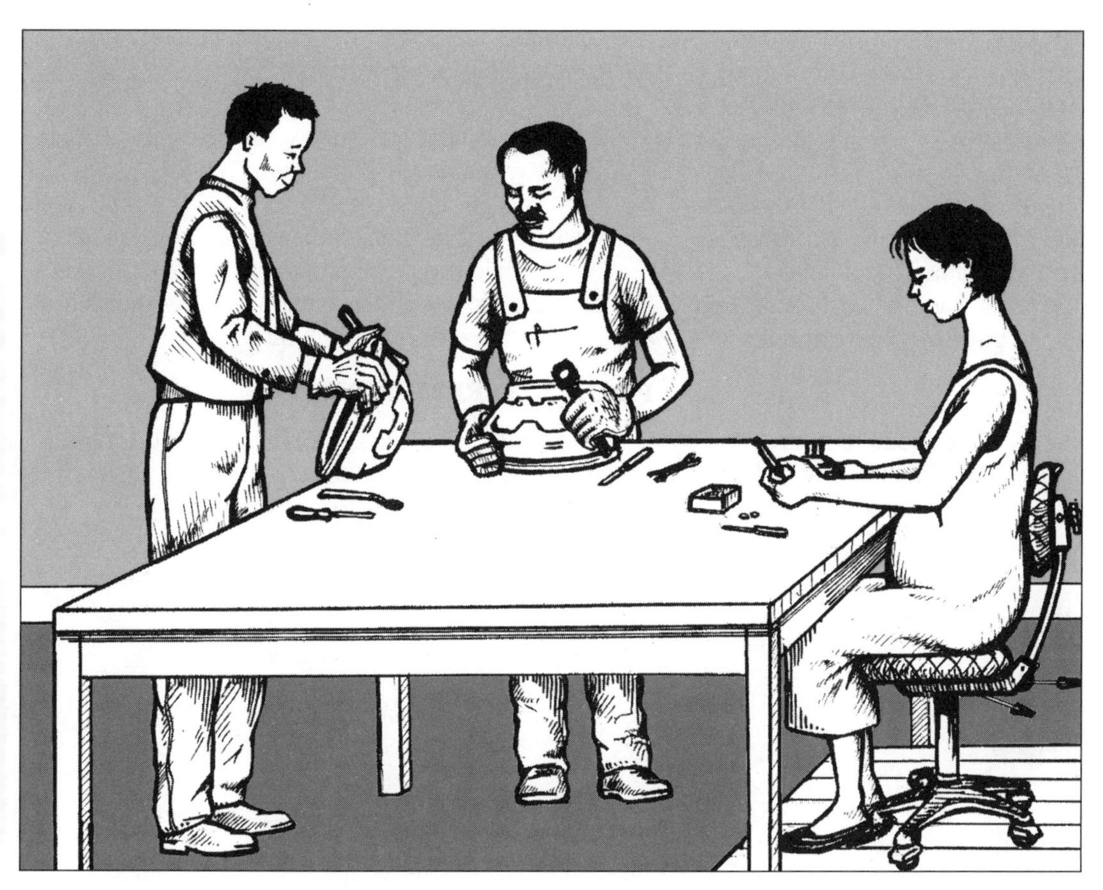

Figure 125a. Make the tasks assigned to pregnant women comfortable and individually adjustable

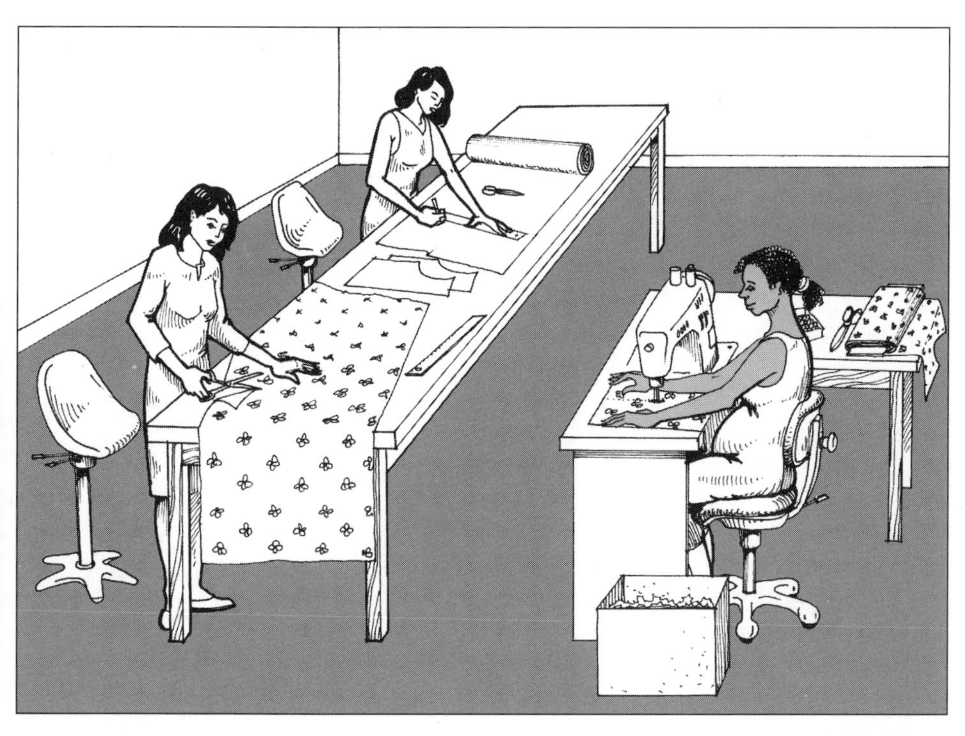

Figure 125b. Do not assign pregnant women to tasks requiring heavy manual work or prolonged standing

CHECKPOINT 126

Take measures so that older workers can perform work safely and efficiently.

WHY

Older workers have knowledge and experience but may have difficulty in adapting to physically demanding tasks or too fast a pace of work. Adapting work to older workers can improve safety and the flow of work.

Older workers often find it difficult to read instructions and signs written in small characters or presented under dim light. Special care should be taken to make them easy to see.

New technology can make jobs easier for older workers, but may make it difficult for them to adapt to it. Although older workers may be very experienced, they need training in newly introduced technology just like younger workers.

HOW

1. Check, together with the workers, if some tasks may cause difficulty or unsafe conditions for older workers. Discuss how these tasks can be made more adaptable to older workers.

2. Apply mechanical devices for physically demanding tasks involving older workers. Make sure that they are able to accomplish new tasks safely.

3. Make instructions, signs and labels easy for older workers to read.

4. Provide sufficient lighting for older workers. Install local lights, if necessary.

5. Make the pace of work variable between younger and older workers so that older workers can cope more easily with it.

6. When introducing new technologies, consult workers to see what measures are needed to adapt them to both younger and older workers.

SOME MORE HINTS

- Aside from mechanization, there are a variety of measures to make tasks physically lighter. For example, improving materials handling can greatly help older workers.

- Provide training for older workers on new tasks in a way that is suited to them.

- Group work in which workers can help each other while the pace of work may vary between individuals is a good solution to solve the difficulties that older workers may have.

POINTS TO REMEMBER

Make full use of older workers' knowledge and experience by adapting work to them. Jobs friendly to older workers are jobs friendly to all.

Figure 126. Check, together with older workers, if some tasks may cause difficulty or unsafe conditions for them

CHECKPOINT 127

Establish emergency plans to ensure correct emergency operations, easy access to facilities and rapid evacuation.

WHY

An emergency may happen at any time. In order to be prepared for it, all those concerned should know in advance what to do in such an emergency. Emergency plans are essential in any enterprise.

Good emergency plans can minimize the consequences of a potential emergency. They can even prevent a serious accident from occurring.

There are priorities for action in any emergency. It is not easy to recall these priorities when you suddenly face an emergency. People need to be instructed in advance and trained repeatedly to respect these priorities for emergency action.

HOW

1. Make a reasonable guess about the nature of potential accidents and identify, by group discussion, types of action that should be taken in each type of emergency. It is especially important to know the likelihood and foreseen consequences of fires, explosions, serious releases of hazardous substances, injuries due to machines and vehicles, and other potential causes of serious injuries such as falls or being struck by objects.

2. Also through group discussion, establish what priority actions should be taken in each type of emergency. These may include emergency operations, shut-down procedures, calling in outside help, first-aid and evacuation methods. This discussion must involve supervisors, workers, and safety and health personnel.

3. Make emergency actions and evacuation procedures known to all people concerned. Train repeatedly those who may engage in emergency operations and first aid. Conduct evacuation drills.

4. Make sure that a list of telephone numbers necessary for emergency action is clearly posted and updated. Confirm with all workers whether they know where this list is located. Also make sure that all on-site first-aid facilities (e.g. emergency treatment equipment, first-aid boxes, means of transport, protective equipment, etc.) and fire extinguishers are clearly marked and located in places that are readily accessible.

SOME MORE HINTS

- It is important to plan in advance and make known who will be in charge of emergency activities.

- When there are major changes in production, machinery and hazardous chemicals used, make sure that these changes are reflected in emergency plans.

- An assessment of risks that may affect the surroundings of the enterprise should be included in the emergency plans.

POINTS TO REMEMBER

Everyone in the workplace should know exactly what to do in an emergency situation. Good emergency plans can prevent serious accidents.

Figure 127. Make emergency action plans with the participation of workers

CHECKPOINT 128

Learn about and share ways to improve your workplace from good examples in your own enterprise or in other enterprises.

WHY

There are many good examples of improvements in your own enterprise or in other enterprises. They represent types of improvement that have been possible under similar local conditions.

The many different problems at the workplace cannot be solved all at once. Progressive improvements are necessary. Here local examples are a good guide as the benefits of improvements are also visible.

Looking at locally achieved good examples, we can learn about and share ways to improve the use of local materials and skills.

HOW

1. Check workplaces in your own enterprise and list good examples showing improved work methods or safe and healthy conditions. Simple, low-cost solutions are particularly important. Examine how these improvements were carried out.

2. Visit other enterprises in your neighbourhood or look at improvement manuals, and learn from good examples.

3. Discuss possible improvements with a group of people. One practical way to do so is through a brainstorming session.

4. Note down, in telegraphic style, kinds of possible improvement that are similar to these good examples and that are relatively inexpensive.

5. Learning from good examples, try to identify feasible solutions. In the discussion, concentrate on solutions which can be carried out immediately, and which are not too idealistic.

SOME MORE HINTS

- Practical training manuals, designed for modern action-oriented training, can also show many good examples of improvements that may be applicable to your local situation.

- Make full use of small-group discussion involving only a few people in finding feasible solutions similar to good examples that you have seen. Looking at slides or video recordings of these good examples will help greatly. Involve in the discussion workers of the particular workplace in question.

- Beginning with simple, low-cost solutions is always a good policy. As people realize that these low-cost solutions are relatively easy to implement, they are encouraged to start joint action. Keep in mind that most ergonomic solutions are simple in nature and not expensive.

POINTS TO REMEMBER

Local good examples have tremendous power to stimulate our thinking. They show what is possible in local conditions. These good examples are found in your and your neighbours' workplaces.

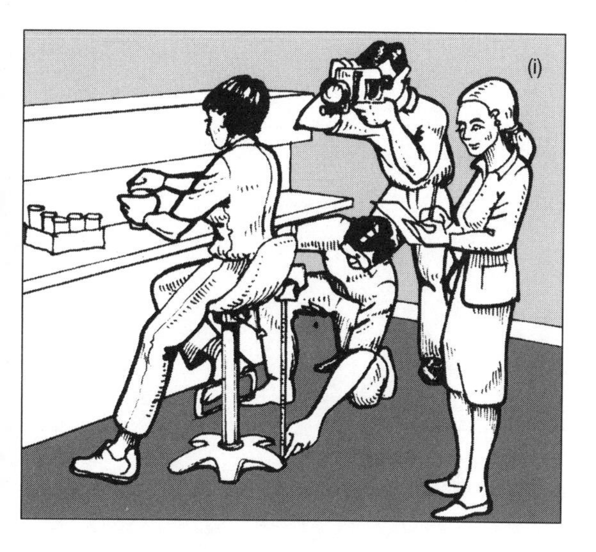

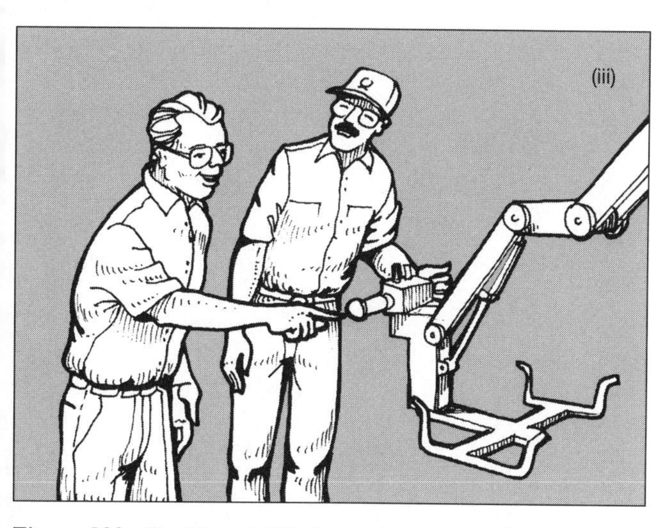

Figure 128. (i), (ii) and (iii) Organize a group (or groups) to check workplaces in your enterprise, learn from good examples and make joint plans for ergonomic improvements